Aircraft Electrical Systems

Other titles by the same author

Aircraft Instruments

Third Edition

Aircraft Electrical Systems

EHJ Pallett

IEng., AMRAeS

Longman
Scientific &
Technical

Copublished in the United States with
John Wiley & Sons, Inc., New York

Longman Scientific & Technical
Longman Group Limited
Longman House, Burnt Mill, Harlow
Essex CM20 2JE, England
and Associated Companies throughout the world

Copublished in the United States with
John Wiley & Sons, Inc., 605 Third Avenue, New York, NY 10158

First published by Pitman Books Limited 1976
Second edition 1979
Third edition 1987
Sixth impression 1994

British Library Cataloguing in Publication Data

Pallett, E. H. J.
    Aircraft electrical systems. —— 3rd ed.
    1. Airplanes —— Electric equipment
    I. Title
    629.135'4   TL690

ISBN 0-582-98819-5

Library of Congress Cataloging in Publication Data

Pallett, E. H. J.
    Aircraft electrical systems.

    Includes index.
    1. Airplanes — Electric equipment.   I. Title.
TL690.P28 1987   629.135'4   86-24553
ISBN 0-470-20734-5 (USA only)

Set in IBM 10/11 pt Press Roman

Produced by Longman Singapore Publishers (Pte) Ltd.
Printed in Singapore

Contents

Preface to the Third Edition

It is now almost eleven years since this book first made its appearance, and the continuing demand warranting this, the third edition, has been most encouraging.

The original sequence of subject structuring has been retained since the reasons noted in the preface to the first edition still apply. It has however, been considered necessary to combine the contents of some chapters, and in others the coverage has been expanded to illustrate the application of principles to a greater number of systems currently in use.

The application of signal processing by means of digital circuit techniques to aircraft systems has been normal practice for a very long time. As far as what may be termed "raw electrical systems" are concerned, the impact of these techniques has, in comparison to such systems as navigation, flight management and automatic flight control, been somewhat less foreboding. However, in relation to those aspects of power generation, distribution and control, it is necessary to have a good understanding of the foregoing techniques, and in particular, the use of logic gates and interpretation of associated diagrams. This latter subject has, therefore, been included in a new chapter to this edition of the book.

In preparing the revised material, the opportunity has been taken to clear up some anomalies that crept into the second edition and subsequent reprints, and I am indebted to those readers who submitted comments. I am also indebted to others who made suggestions regarding the inclusion of new material, and who supplied information for reference purposes.

In conclusion, it is perhaps pertinent to note that this edition has been prepared during a transition from one publisher to another, and so I would like to thank the one under whose logo it now appears, not only for having undertaken their particular tasks, but also for establishing a new publisher/author association.

Copthorne
Sussex
1986

E. P.

Preface to the First Edition

Increases in size and speed, changes in shape and functional requirements of aircraft have each been possible by technical research and development and the progress made not only applies to those visible structural parts, but also to those unseen systems and services which enable it to function as an integrated machine.

A system ranking very highly indeed in this progression is the one concerned with electrical power involving as it does various methods of generation, distribution, control, protection and utilization. These methods do, in fact, form a natural "build-up" of an aircraft's electrical system and their sequence sets a convenient pattern on which a study of principles and applications can be based. The material for this book therefore follows this pattern.

In the early days of what is familiarly called "aircraft electrics", there was a certain distrust of the equipment. Although there was acceptance of the fact that electricity was necessary for operating the "wireless" equipment, a few lights and an engine ignition system, many individuals were inclined to the view that if other systems could not be operated either by air, hydraulic oil, cables, numerous mechanical linkages or petrol, then they were quite unnecessary! A majority of the individuals were mechanics, and the ground engineers as they were then known, and undoubtedly, when "electrickery" began proving itself as a system operating media, it came as a pleasant relief to leave all relevant work to that odd character, the electrician, who speaking in some strange jargon and by means of diagrams containing numerous mystic lines and symbols, seemed better able to cope with it all!

With the continued development of the various types of aircraft, the sources of electrical power have also varied from the simple battery and wind-driven generator, through to the most complex multiple a.c. generating systems. Similarly, the application of power sources have varied and in conjunction with developments in electronics, has spread into the areas of other systems to the extent of performing not only a controlling function but, as is now so often the case, the entire operating function of a system. As a result, the work of the electrician assumed greater importance and has become highly specialized, while other maintenance specialists found, and continue to find it increasingly necessary to broaden their knowledge of the subject; indeed it is incumbent on them to do so in order to carry out their important duties. This also applies to pilots in order that they may meet the technical knowledge requirements appropriate to their duties and to the types of aircraft they fly.

Fundamental electrical principles are described in many standard text books, and in preparing the material for this book it was in no way intended that it should supplant their educational role. However, it has been considered convenient to briefly review certain relevant principles in the chapters on generation and conversion of power supplies, to "lead-in" to the subject and, it is hoped, to convey more clearly how they are applied to the systems described. In keeping with the introductory nature of the book, and perhaps more important, to keep within certain size limitations, it obviously has not been possible to cover all types of aircraft systems. However, in drawing comparisons it is found that applications do have quite a lot in common, and so the examples finally chosen may be considered sufficiently representative to provide a useful foundation for further specialized study.

The details given embrace relevant sections of the various syllabuses established for the technical examination of maintenance engineers and pilots by official organizations, training schools and professional societies. In this connection, therefore, it is also hoped that the book will provide a useful source of reference.

A selection of questions are provided for each chapter and the author is indebted to the Society of Licensed Aircraft Engineers and Technologists for permission to reproduce questions selected from examination papers.

Valuable assistance has been given by a number of organizations in supplying technical data, and in granting permission to reproduce many of the illustrations, grateful acknowledgement is hereby made to the following —
Amphenol Ltd.
Auto Diesels Braby Ltd.
Aviquipo of Britain Ltd.
Belling & Lee Ltd.
B.I.C.C.
British Aircraft Corporation (Operating) Ltd.
Britten-Norman Ltd.
Cannon Electric (G.B.) Ltd.
Davall.
Dowty Electrics Ltd.
Graviner (Colnbrook) Ltd.

Hawker Siddeley Aviation Ltd.
Honeywell Ltd.
International Rectifier Co. (G.B.) Ltd.
Lucas Aerospace Ltd.
Newton Brothers (Derby) Ltd.
Normalair-Garrett Ltd.
Plessey Co., Ltd.
SAFT (United Kingdom) Ltd.
Sangamo Weston Ltd.
Shell Aviation News.
Smiths Industries Ltd.
Standard Telephones & Cables Ltd.
Thorn Bendix.
Varley Dry Accumulators Ltd.

Finally, thanks are also due to the publishers for having patiently awaited the completion of sections of manuscript and also for having accepted a number of changes of subject.

Copthorne,   E.P.
Sussex

CHAPTER ONE

Direct Current Power Supplies

INTRODUCTION

Depending on the type of aircraft, and the extent to which electrical power is to be utilized for the operation of its systems and components, the primary supply of such power may either be direct current (d.c.) or alternating current (a.c.). This chapter deals with the first of these supplies and how it is produced by both generators and batteries. Examples of some typical aircraft systems are described in Appendix 7.

Fundamental Principles of Generators

A generator is a machine that converts mechanical energy into electrical energy by the process of electromagnetic induction. In both d.c. and a.c. types of generator, the voltage induced is alternating; the major difference between them being in the method by which the electrical energy is collected and applied to the circuit externally connected to the generator.

Figure 1.1(a) illustrates a generator in its simplest form, i.e. a single loop of wire "AB" arranged to rotate between the pole pieces of a magnet. The ends of the wire are brought together to form a circuit via slip rings, brushes and the externally connected load. When the plane of the loop lies at right angles to the magnetic field (position 1, Fig. 1.1(b)) no voltage is induced in the loop. As the loop rotates through 90 degrees the wires cut the lines of force at right angles until at position 2 the induced voltage is at a maximum. As the loop approaches the vertical position again the voltage decreases since the rate at which lines of force are cut diminishes. At position 3 the induced voltage is zero. If rotation is continued, the number of lines cut gradually increases, until at 270 degrees (position 4) it is once again maximum, but as the cutting is in the opposite direction there is also

a reversal of the direction of induced voltage. As rotation continues, the number of lines cut decreases and the induced voltage reduces to zero as the loop returns to position 1. Plotting of the induced voltage throughout the full cycle produces the alternating or sine curve shown.

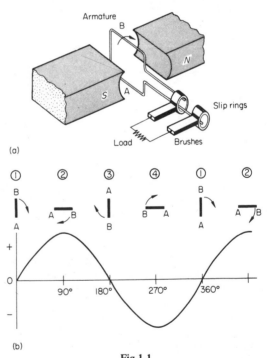

Fig 1.1
(a) Simple form of generator
(b) Induced voltage

To convert the a.c. produced into unidirectional or d.c., it is necessary to replace the slip rings by a collecting device referred to as a commutator. This is shown in Fig. 1.2 (a) and as will be noted it consists of two segments insulated from each other and connected to the ends of the loop. The brushes are set so that each segment moves out of contact with one brush and into contact with the other at the point where the loop passes through the positions at which induced voltage is minimum. In other words, a pulsating current increasing to maximum in one direction only is produced as shown by the curve in Fig. 1.2(b).

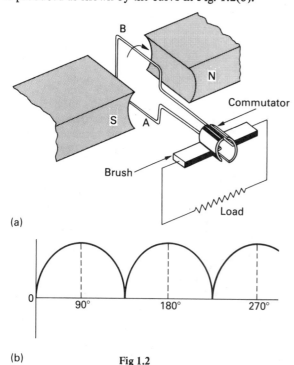

(a)

(b)

Fig 1.2
Conversion of a.c. to d.c.
(a) Use of commutator
(b) Current wave-form

In order to smooth out the pulsations and to produce a more constant output, additional wire loops and commutator segments are provided. They are so interconnected and spaced about the axis of rotation, that several are always in a position of maximum action, and the pulsating output is reduced to a ripple as indicated in Fig. 1.3.

Generator Classifications

Generators are classified according to the method by which their magnetic circuits are energized, and the following three classes are normally recognized —
(1) Permanent magnet generators.

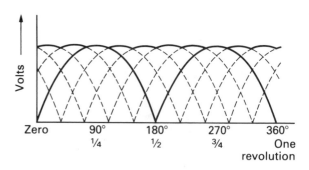

Fig 1.3
Effect on output using several coils

(2) Separately-excited generators, in which electromagnets are excited by current obtained from a separate source of d.c.
(3) Self-excited generators, in which electromagnets are excited by current produced by the machines themselves. These generators are further classified by the manner in which the fixed windings, i.e. the electromagnetic field and armature windings, are interconnected.

In aircraft d.c. power supply systems, self-excited shunt-wound generators are employed and the following details are therefore related only to this type.

Fixed Winding Arrangement

Figure 1.4 illustrates the arrangement of the fixed windings of a basic four-pole machine suitable for use as a self-excited generator. The fixed portion of the armature circuit consists of the four brushes, the links connecting together brushes of like polarity and the

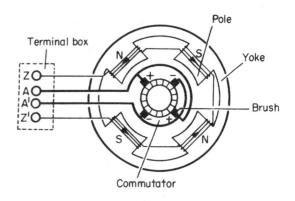

Fig 1.4
Fixed winding arrangements

cables connecting the linked brushes to the terminals indicated A and A^1. The four field coils are of high resistance and connected in series to form the field winding. They are wound and connected in such a way that they produce alternate North and South polarities. The ends of the windings are brought out to the terminals indicated as Z and Z^1.

Generator Characteristics

The characteristics of a generator refer to the relationship between voltage and the current flowing in the external circuit connected to a generator, i.e. the load current, and there are two which may be closely defined. These are: the *external* characteristic or relationship between *terminal voltage and load current,* and the *internal* characteristic or relationship between the *actual electromagnetic force (e.m.f.) generated in the armature windings and load current.* These relationships are generally shown in the form of graphs, with the graph drawn for one particular speed of the generator.

Self-excited Shunt-wound Generators

Shunt-wound generators are one of three types in the self-excited class of machine and as already noted are used in aircraft d.c. power supply systems. The term "shunt-wound" is derived from the fact that the high-resistance field winding is connected across or in parallel with the armature as shown in Fig. 1.5. The

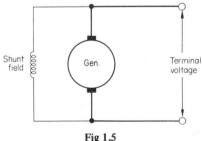

Fig 1.5
Connection of shunt-field winding

armature current divides into two branches, one formed by the field winding, the other by the external circuit. Since the field winding is of high resistance, the advantage is gained of having maximum current flow through the external circuit and expenditure of unnecessary electrical energy within the generator is avoided.

Operating Principle and Characteristic

When the armature is rotated the conductors cut the weak magnetic field which is due to residual magnetism in the electromagnet system. A small e.m.f. is induced in the armature winding and is applied to the field winding, causing current to flow through it and so increasing the magnetic flux. This, in turn, causes a progressive increase in the induced e.m.f. and field current until the induced e.m.f. and terminal voltage reach the steady open-circuit maximum.

The characteristic for this type of generator is shown in Fig. 1.6 and it will be observed that the terminal

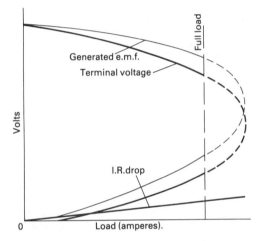

Fig 1.6
Characteristic of self-excited shunt-wound generator

voltage tends to fall with increasing load current. This is due to the voltage drop (IR drop) in the armature winding and also to a weakening of the main flux by armature reaction. The fall in terminal voltage reduces the field current, the main flux is further weakened and therefore a further fall in terminal voltage is produced.

If the process of increasing the load is continued after the full working load condition has been reached, the terminal voltage will fall at an increasing rate until it can no longer sustain the load current and they both fall to zero. With reduced excitation the external characteristic of a shunt-wound generator falls much more rapidly so that the point at which voltage collapse occurs will be reached with a much smaller load current. In practice, field current is adjusted to maintain constant voltage under all load conditions, by a voltage regulator the operation of which will be described later.

Sometimes a generator will lose its residual magnetism or become incorrectly polarized because of heat, shock, or a momentary current in the wrong direction. This can be corrected by momentarily passing current through the field from the positive terminal to the negative terminal; a procedure known as "flashing the field".

Generator Construction

A typical self-excited shunt-wound four-pole generator, which is employed in a current type of turbo-prop civil transport aircraft, is illustrated in Fig. 1.7. It is designed to provide an output of 9 kilowatts at a continuous current of 300 amperes (A) over the speed range of 4,500 to 8,500 rev/min. In its basic form the construction follows the pattern conventionally adopted and consists of five principal assemblies; namely, the yoke, armature, two end frames and brush-gear assembly.

THE YOKE

The yoke forms the main housing of the generator, and is designed to carry the electromagnet system made up of the four field windings and pole pieces. It also provides for the attachment of the end frame assemblies. The windings are pre-formed coils of the required ampere-turns, wound and connected in series in such a manner that when mounted on the pole pieces, the polarity of the field produced at the poles by the coil current is alternately North and South (see Fig. 1.4). The field windings are suitably insulated and are a close fit on the pole pieces which are bolted to the yoke. The faces of the pole pieces are subjected to varying magnetic fields caused by rotation of the armature, giving rise to induced e.m.f. which in turn produces eddy currents through the pole pieces causing local heating and power wastage. To minimize these effects the pole pieces are of laminated construction; the thin soft iron laminations being oxidized to insulate and to offer high electrical resistance to the induced e.m.f.

INTERPOLE AND COMPENSATING WINDINGS

During operation on load, the current flowing through the armature winding of a generator creates a magnetic field which is superimposed on the main field produced by field-winding current. Since lines of force cannot intersect, the armature field distorts the main field by an amount which varies with the load; such distorting effect is termed armature reaction. If

uncorrected, armature reaction produces two additional undesirable effects: (i) it causes a shift of the Magnetic Neutral Axis, i.e. the axis passing through two points at which no e.m.f. is induced in a coil, setting up reactive sparking at the commutator, and (ii) it weakens the main field causing a reduction in generated e.m.f. The position of the brushes can be altered to minimize these effects under varying load conditions, but a more effective method is to provide additional windings in the electromagnet system, such windings being referred to as interpole and compensating windings.

Interpole windings are wound on narrow-faced auxiliary pole pieces located midway between the main poles, and are connected in series with the armature. The windings are such that an interpole has the same polarity as the next main pole in the direction of rotation, and as the fluxes are opposite in direction to the armature flux, they can be equalized at all loads by having the requisite number of turns.

In order to provide true correction of armature reaction, the effects produced by interpoles must be supplemented, since alone they cannot entirely eliminate all distortion occurring at the main pole faces. Compensating windings are therefore connected in series with the interpole and armature windings, and located in slots cut in the faces of the main pole shoes. The sides of the coils thus lie parallel with the sides of the armature coils. The ampere-turns of the winding are equal to those of the armature winding, while the flux due to it is opposite in direction to the armature flux.

AUXILIARY INTERPOLES

The effectiveness of interpoles in minimizing reactance sparking is limited by armature speed, and their application as individual components of a field-winding system is, therefore, restricted to generators operating over a narrow speed range, e.g. the designed range of the generator illustrated in Fig. 1.7. In the case of generators designed for operation over a wide range, e.g. 2850 rev/min up to 10,000 rev/min, the use of interpoles alone would produce a side effect resulting in reactance sparking as the generator speed is reduced from maximum to minimum. To counteract this, and for a given load on the generator, it is necessary to reduce the magnetomotive force (m.m.f.) of the interpoles. The desired effect may be obtained by winding auxiliary coils over the interpole coils and connecting them in series with the generator shunt field winding in such a way that each coil, when energized by shunt field circuit current, produces an

5

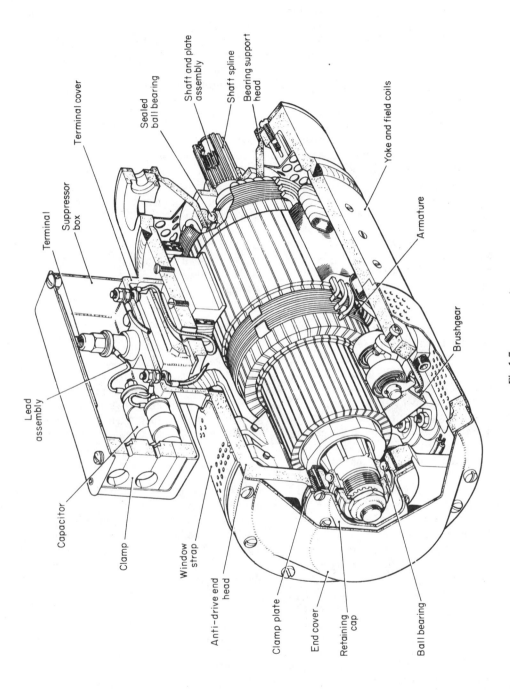

Terminal cover

Shaft and plate assembly

Sealed ball bearing

Shaft spline

Bearing support head

Terminal

Suppressor box

Yoke and field coils

Lead assembly

Armature

Capacitor

Clamp

Brushgear

Window strap

Anti-drive end head

Clamp plate

End cover

Retaining cap

Ball bearing

Fig 1.7
Sectioned view of a generator

m.m.f. of opposite polarity to that produced by the interpole coil on the same pole shoe. An exact balance between reactance e.m.f. and commutation e.m.f. is maintained over the full working range of generator speed to assist in producing sparkless commutation.

ARMATURE ASSEMBLY

The armature assembly comprises the main shaft (which may be solid or hollow) core and main winding, commutator and bearings; the whole assembly being statically and dynamically balanced. In the generator shown, the shaft is hollow and internally splined to mate with splines of a drive shaft which passes through the entire length of the armature shaft.

Armature windings are made up of a number of individual identical coils which fit into slots at the outer edges of steel laminations which form the core of the armature. The coils are made from copper strip and as security against displacement by centrifugal force, steel wire (in some cases steel strip) is bound round the circumference of the armature. The ends of each coil are brought out to the commutator and silver brazed to separate segments, the finish of one coil being connected to the same segment as the beginning of another coil. The complete winding thus forms a closed circuit. The windings are invariably vacuum-impregnated with silicone varnish to maintain insulation resistance under all conditions.

In common with most aircraft generators, the commutator is of small diameter to minimize centrifugal stressing, and is built up of long, narrow copper segments corresponding in number to that of the field coils (a typical figure is 51 coils). The segment surfaces are swept by brushes which are narrow and mounted in pairs (usually four pairs) to maintain the brush contact area per segment — an essential prerequisite for effective commutation.

The armatures of all aircraft generators are supported in high efficiency ball or roller bearings, or in combinations of these two types. Where combinations are used in a single generator it will be found that the ball bearing is invariably fitted at the drive end of the armature shaft, and the roller bearing at the commutator end. This arrangement permits lateral expansion of the armature shaft, arising from temperature increases in the generator, without exposing the bearings to risk of damage. Bearings are lubricated either with a specified high-melting-point grease or lubricating oil and may be of the sealed or non-sealed types. Sealed grease-lubricated bearings are pre-packed by the manufacturer and require no further

lubrication during the life of the bearing. Non-sealed grease-lubricated bearings are assembled with sufficient lubricant to last for the period of the generator servicing cycle. In general the lubricant for oil-lubricated bearings is introduced into the bearing through the medium of oil-impregnated felt pads. Seals are provided to prevent oil escaping into the interior of the generator.

END FRAME ASSEMBLIES

These assemblies are bolted one at each end of the yoke and house the armature shaft bearings. The drive end frame provides for the attachment of the generator to the mounting pad of the engine or gear-box drive (see also p. 8) and the commutator and frame provides a mounting for the brush-gear assembly and, in the majority of cases, also provides for the attachment of a cooling air duct. Inspection and replacement of brushes is accomplished by removing a strap which normally covers apertures in the commutator end frame.

BRUSH-GEAR ASSEMBLY

The brush-gear assembly is comprised of the brushes and the holding equipment necessary for retaining the brushes in the correct position, and at the correct angle with respect to the magnetic neutral axis.

Brushes used in aircraft generators are of the electro-graphitic type made from artificial graphite. The graphite is produced by taking several forms of natural carbons, grinding them into fine powder, blending them together and consolidating the mixture into the desired solid shape by mechanical pressure followed by exposure to very high temperature in an electric furnace. These brushes possess both the robustness of carbon and the lubricating properties of graphite. In addition they are very resistant to burning by sparking, they cause little commutator wear and their heat conductivity enables them to withstand overloads.

As stated earlier, an essential prerequisite for effective commutation is that brush contact area per commutator segment should be maintained. This is accomplished by mounting several pairs of brushes in brush holders; in the generator illustrated in Fig. 1.7 four pairs of brushes are employed. The holders take the form of open-ended boxes whose inside surfaces are machined to the size of a brush, plus a slight clearance enabling a brush to slide freely without tilting or rocking. Contact between brushes and commutator is maintained by the pressure exerted by the free ends of adjustable springs anchored to posts on the brush holders. Springs are adversely affected by current passing through them; it is usual, therefore, to fit an insulating pad or roller at the end of the spring where it bears on the top surface of the brush.

The brush holders are secured either by bolting them to a support ring (usually called a brush rocker) which is, in turn, bolted to the commutator end frame, or as in the case of the generator illustrated, bolted directly to the end frame. In order to achieve the best possible commutation a support ring, or end frame, as appropriate, can be rotated through a few degrees to alter the position of the brushes relative to the magnetic neutral axis. Marks are provided on each generator to indicate the normal operating position.

When four or more brush holders are provided, they are located diametrically opposite and their brushes are alternately positive and negative, those of similar polarity being connected together by bar and flexible wire type links.

The brushes are fitted with short leads or "pigtails" of flexible copper braid moulded into the brush during manufacture. The free ends of the pigtails terminate in spade or plate type terminals which are connected to the appropriate main terminals of the generator via the brush holders and connecting links.

TERMINAL BLOCKS
The leads from brush-gear assemblies and field windings are connected to terminal posts secured to a block mounted on the commutator end frame or, in some generators, on the yoke assembly (see Fig. 1.7). The terminals and block are enclosed in a box-like cover also secured to the end frame. Entry for the output supply cables of the distribution system (refer to Chapter 5) is through rubber clamps. The rotation of a generator armature is specified in a direction, normally anti-clockwise, when viewed from the drive end assembly. A movable link is fitted between two of the terminals which can be connected in an alternative position should it be necessary for the generator to be driven in the reverse direction.

SPARK SUPPRESSION
Sparking at the brushes of a generator, no matter how slight, results in the propagation of electromagnetic waves which interfere with the reception of radio signals. The interference originating in generators may be eliminated quite effectively by screening and suppression. Screening involves the enclosure of a generator in a continuous metallic casing and the sheathing of output supply cables in continuous metallic tubing or conduit to prevent direct radiation. To prevent interference being conducted along the distribution cable system, the screened output supply cables are terminated in filter or suppressor units. These units consist of chokes and capacitors of suitable electrical rating built into metal cases located as close to a generator as possible. Independent suppressor units are rather cumbersome and quite heavy, and it is therefore the practice in the design of current types of generator to incorporate internal suppression systems. These systems do not normally contain chokes, but consist simply of suitably rated capacitors (see Fig. 1.7) which are connected between generator casing (earth) and terminals. The use of internal suppression systems eliminates the necessity for screened output supply cables and conduits thereby making for a considerable saving in the overall weight of a generator installation.

Rectified Power Supplies

In many of the smaller types of single-engined and twin-engined aircraft, the primary d.c. power is supplied in a manner similar to that of automobiles, i.e. it is a rectified output from a frequency-wild alternating current generator. Its operating frequency is about 100 Hz at idling speed of the engine and increases with speed to 1200 Hz or higher.

The generator or alternator as it is more generally called, consists of a rotor, stator, slip ring and brush assembly and end frames. In addition six silicon diodes are carried in an end frame and are connected as a bridge rectifier (see p. 57) to provide the d.c. for the aircraft's system. The principal constructional features are illustrated in Fig. 1.8.

The rotor is formed by two extruded steel pole pieces which are press-fitted on to the rotor shaft to

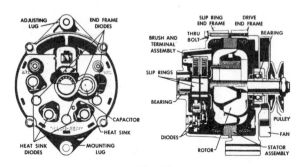

Fig. 1.8
Alternator supplying a rectified output

sandwich a field coil and thus form the core of the electromagnet. Each pole piece has six "fingers" which in position, mesh but do not touch each other. Excitation current is fed to a field coil on the rotor via brushes, and slip-rings which are press-fitted onto the rotor shaft.

The stator is made up of a number of steel stampings riveted together to form the core around which the three star-connected phase coils are wound. One end of each winding is connected to the bridge rectifier assembly while the other ends are joined to form what is termed the neutral point. The stator assembly is clamped between the end frames.

Figure 1.9 illustrates the circuit diagram of the alternator. Unlike a conventional d.c. generator, the alternator has no residual magnetism and so its field must be excited initially by d.c. from the aircraft's battery or an external power supply. When d.c. is switched on to the generator, the rotor field coil is energized and the pole piece "fingers" become alternately north and south magnetic poles. As the rotor rotates, the field induces a three-phase alternating current within the stator which is fed to the diodes for rectification, and then to the aircraft's system. As will be noted from Fig. 1.9,

when the alternator is supplying the busbar, it will also supply its own field excitation current to sustain the regulated output. The level of voltage is regulated by a solid-state type of voltage regulator (see p. 14).

Generator/Engine Coupling

Depending on the type and application, a generator may be driven by an engine either from an accessories gear box, or by a pulley and belt.

The generator already shown in Fig. 1.7, is an example of one driven through gearing which forms part of an accessories gear-box. Depending on the rated output of a generator and on the load requirements of the electrical system of a particular aircraft, there is a specific gear drive ratio.

The drive from the gear-box is by means of a quill shaft with either male or female serrations or splines at one or both ends. The serrations or splines mate with corresponding formations on the generator armature shaft (see Fig. 1.7) to transmit the torque delivered by the driving gear. One of the requirements to be satisfied by a quill drive is that it must effectively interrupt transmission of the driving torque in

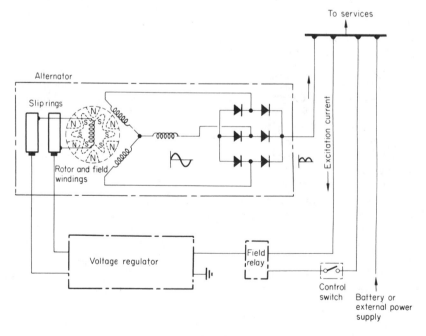

Fig 1.9
Circuit diagram of alternator

the event that the generator armature seizes up. This is done by designing the drive shaft so that at one section its diameter is smaller than the remaining sections; thus providing a weak spot at which the shaft will shear under the effect of an excessive torque.

Quill drives are usually short and rigid, but in some cases a long drive with one end mating with serrations formed deep in a hollow armature shaft may be specified. This arrangement enables the drive to absorb much of the mechanical vibration which is otherwise passed to a generator from an accessories gear-box.

The method of securing a generator to an accessories gear-box varies, but in general it is either one utilizing a mounting flange or one requiring a manacle ring. In the mounting flange method, the end frame at the drive end of a generator is usually extended to a larger diameter than the yoke, thus forming a projecting flange. Holes in the flange line up with and accept studs which are located in the mounting pad of the engine or gear-box, and the generator is finally secured by nuts, locking washers, etc. An alternative form of flange mounting is based on a generator end frame having two diameters. The larger diameter is no greater than that of the yoke and abuts on the mounting pad while the reduced diameter provides a channel or "gutter", between the yoke and the larger diameter of the end frame, into which the mounting studs project. Another variation of this form of mounting is employed in the generator shown in Fig. 1.7.

In the manacle-ring method of mounting the generator drive end frame has an extension with a recess in the mounting face of the driving unit. When the generator extension is fully engaged with the recess, a flange on the end frame abuts on a matching flange formed on the driving unit mounting face. The two flanges are then clamped together by a manacle ring which, after being placed over them, is firmly closed by a tensioning screw. A spigot arrangement is usually incorporated to provide location of the generator to the drive unit, and to absorb torque reaction when the generator is operating.

The pulley and belt drive is commonly adopted for driving alternators of the type shown in Fig. 1.8, and as may be seen from Fig. 1.10, it is similar in many respects to the one adopted in automobiles.

The alternator is secured to two mounting brackets one of which is slotted, so that when the corresponding securing bolt is slackened, the alternator may be positioned about the other bolt for

the purpose of adjusting belt tension. The required drive ratio is, of course, determined by the diameters of the engine and alternator pulleys.

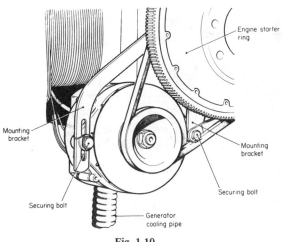

Fig 1.10
Pulley and belt drive

Cooling of Generators

The maximum output of a generator, assuming no limit to input mechanical power, is largely determined by the ease with which heat (arising from hysteresis, thermal effect of current in windings, etc.) can be dissipated. With large-bulk generators of relatively low output the natural processes of heat radiation from the extensive surfaces of the machine carcase may well provide sufficient cooling, but such "natural" cooling is inadequate for the smaller high-output generators used for the supply of electrical power to aircraft, and must, therefore, be supplemented by forced cooling.

The most commonly accepted method of cooling is that which utilizes the ram or blast effect resulting from either the slipstream of a propeller or the airstream due to the aircraft's movement. A typical cooling system is shown in a basic form in Fig. 1.11. The air is forced at high speed into an intake and is led through light-alloy ducts to a collector at the commutator end of the generator. The air discharges over the brush-gear and commutator to cool this natural area of high temperature, and then passes through the length of the machine to exhaust through apertures, surrounded by a perforated strap, at the drive end. In order to assist in ram-air cooling and also to provide some cooling when the aircraft is on the ground, many types of generator have a fan fitted at the drive end of the armature shaft.

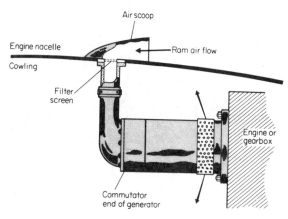

Fig. 1.11
Typical cooling system

Cooling of the alternator shown in Fig. 1.8 is provided by a fan at the driving end and by air passing through slotted vents in the slip ring end frame. Heat at the silicon diodes is dissipated by mounting them on steel plates known as "heat sinks".

Brush Wear

The carbon from which electro-graphitic brushes are made is extremely porous and some of the pores are so very fine that carbon has an exceptional ability to absorb other substances into its structure, and to retain them. Moisture is one of these substances and it plays an important part in the functioning of a brush contact by affording a substantial degree of lubrication. The moisture is trapped under the inevitable irregularities of the contact faces of the brushes and forms an outside film on the commutator and it is with this film that the brushes make contact. Just how vital a part moisture does play was, however, not fully realized until aircraft began operating at high altitudes and the problem arose of brushes wearing out very rapidly under these conditions. Investigations into the problem showed that the fundamental difficulty was the extreme dryness of the atmosphere, this, in its turn, producing three secondary effects: (i) friction between brushes and commutator because the lubricating film cannot form, (ii) contact resistance becomes negligible giving rise to heavy reactive sparking and accelerated brush erosion and (iii) static electrical charges due to friction, producing molecular breakdown of the brushes.

These effects have been largely eliminated by using brushes which have a chemical additive as a means of replacing the function which atmospheric moisture plays in surface skin formation. Two distinct categories are in general use: brushes of one category form a constant-resistance semi-lubricating film on the commutator, while those in the other category are, in effect, self-lubricating brushes which do not form a film.

The composition of the film-forming brushes includes chemicals (e.g. barium fluoride) to build up progressively a constant-resistance semi-lubricating film on the commutator surfaces. Brushes of this category do not wear abnormally at altitudes up to 60,000 feet providing that generators to which they are fitted have been previously "bench run" for some hours to allow the formation of the protective film. This film, once formed, is very dark in colour and may often give the impression of a dirty commutator.

Brushes of the non-film-forming category contain a lubricating ingredient such as molybdenum disulphide which is often packed in cores running longitudinally through the brushes. Since the brush is self-lubricating it is unnecessary for generators fitted with this type to be run for hours prior to entering service. However, they do have the disadvantage of appreciably shorter life, due to somewhat more rapid wear, when compared with film-forming brushes.

Voltage Regulation

The efficient operation of aircraft electrical equipment requiring d.c. depends on the fundamental requirement that the generator voltage at the distribution busbar system be maintained constant under all conditions of load and at varying speeds, within the limits of a prescribed range. It is necessary, therefore, to provide a device that will regulate the output voltage of a generator at the designed value and within a specified tolerance.

There are a number of factors which, either separately or in combination, affect the output voltage of a d.c. generator, and of these the one which can most conveniently be controlled is the field circuit current, which in its turn controls the flux density. This control can be effected by incorporating a variable resistor in series with the field winding as shown in Fig. 1.12. Adjustments to this resistor would vary the resistance of the field winding, and the field current and output voltage would also vary and be brought to the required controlling value. The application of the resistor in the manner indicated is, however, limited since it is essential to incorporate a regulating device which will automatically respond to changes of load and speed, and also, automatically make the necessary adjustments to the generator field current. Three of the regulation methods commonly adopted are: the vibrating contact method; the one based on the pressure/resistance characteristics of carbon, namely, the carbon pile method, and the one based on solid-state circuit principles.

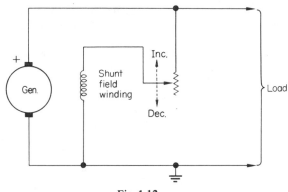

Fig 1.12
Control of field circuit current

Vibrating Contact Regulator   Vibrating contact regulators are used in several types of small aircraft employing comparatively low d.c. output generators and a typical circuit for the regulation of both voltage and current of a single generator system is shown in basic form in Fig. 1.13. Although the coil windings of each regulator are interconnected, the circuit arrangement is such that either the voltage regulator only or the current regulator only can operate at any one time. A third unit, called a reverse current cut-out relay, also forms part of some types of regulator, and since the relay has a circuit protection function, a description of its construction and operation will be given in Chapter 7.

Voltage Regulator   This unit consists of two windings assembled on a common core. The shunt winding consists of many turns of fine gauge wire and is connected in series with the current regulator winding and in parallel with the generator. The series winding, on the other hand, consists of a few turns of heavy gauge wire, and is connected in series with the generator shunt-field winding when the contacts of both regulators are closed, i.e. under static condition of the generator system. The contact assembly is comprised of a fixed contact and a movable contact secured to a flexibly-

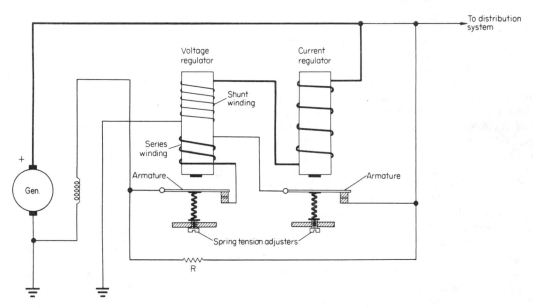

Fig 1.13
Vibrating contact regulator principle

hinged armature. Movement of the armature and, therefore, the point at which contact opening and closing takes place is controlled by a spring which is pre-adjusted to the required voltage setting.

When the generator starts operating, the contacts of both regulators remain closed so that a positive supply can flow through the generator shunt-field winding to provide the necessary excitation for raising the generator output. At the same time current passes through the shunt winding of the voltage regulator and, in conjunction with the series winding, it increases the regulator's electromagnetic field. As soon as the generator output voltage reaches the pre-adjusted regulator setting, the electromagnetic field becomes strong enough to oppose the tension of the armature spring thereby opening the contacts. In this equilibrium position, the circuit to the series winding is opened causing its field to collapse. At the same time, the supply to the generator field winding passes through a resistance (R) which reduces the excitation current and, therefore, the generator output voltage. The reduced output in turn reduces the magnetic strength of the regulator shunt winding so that spring tension closes the contacts again to restore the generator output voltage to its regulated value and to cause the foregoing operating cycle to be repeated. The frequency of operation depends on the electrical load carried by the generator; a typical range is between 50 to 200 times a second.

In regulators designed for use with twin-generator systems, a third coil is also wound on the electromagnet core for paralleling purposes (see p. 16) and is connected to separate paralleling relays.

Current Regulator   This unit limits generator current output in exactly the same way as the voltage regulator controls voltage output, i.e. by controlling generator field-excitation current. Its construction differs only by virtue of having a single winding of a few turns of heavy wire.

When electrical load demands are heavy, the voltage output value of the generator may not increase sufficiently to cause the voltage regulator to open its contacts. Consequently, the output will continue to increase until it reaches rated maximum current, this being the value for which the current regulator is set. At this setting, the current flowing through the regulator winding establishes a strong enough electromagnetic field to attract the armature and so open the contacts. Thus, it is the current regulator which now inserts resistance R in the generator shunt-field circuit

to reduce generator output. As soon as there is sufficient drop in output the field produced by the regulator winding is overcome by spring tension, the contacts close and the cycle again repeated at a frequency similar to that of the voltage regulator.

Carbon Pile Regulator   Carbon has a granular surface and the contact resistance between two carbon faces that are held together depends not only on the actual area of contact, but also on the pressure with which the two faces are held together. If, therefore, a number of carbon discs or washers are arranged in the form of a pile and connected in series with the shunt field of a generator (see Fig. 1.14) the field circuit resistance can be varied by increasing or decreasing the pressure applied to the ends of the pile and changes in generator output voltage therefore counteracted. Since this method eliminates the use of vibrating contacts, it is applied to generators capable of high current output, and requiring higher field excitation current. The necessary variation of pile pressure or compression under varying conditions of generator speed and load, is made through the medium of an electromagnet and spring-controlled armature which operate in a similar manner to those of a vibrating contact regulator.

Under static conditions of the generator system, the carbon pile is fully compressed and since there is no magnetic "pull" on the armature, the resistance in the generator shunt-field circuit is minimum and the air gap between the regulator armature and electromagnet core is maximum. As the generator starts operating, the progressively increasing output voltage is applied to the regulator coil and the resulting field establishes an increasing "pull" on the armature. During the initial "run-up" stages, the combination of low voltage applied to the regulator coil, and the maximum air gap between armature and core, results in a very weak force of attraction being exerted on the armature. This force is far smaller than that of the spring control, hence the armature maintains its original position and continues to hold the carbon pile in the fully compressed condition; the shunt-field circuit resistance is thus maintained at minimum value during run-up to allow generator output voltage to build up as rapidly as possible. This condition continues unaltered until the voltage has risen to the regulated value, and at which equilibrium is established between magnetic force and spring-control force. The armature is free to move towards the electromagnet core if the force of magnetic attraction is increased as a result of any increase in generator speed within the effective

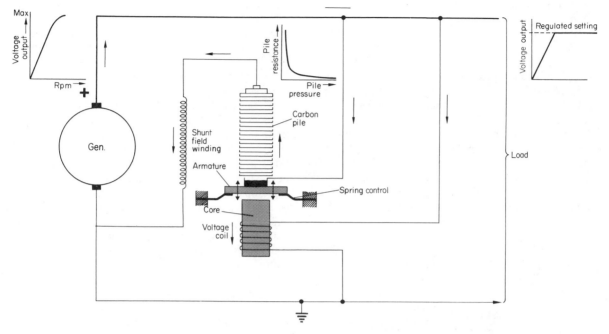

Fig 1.14
Carbon pile voltage regulation

speed range. In these circumstances pile compression is further reduced so that there is more air space between discs to increase resistance and so check a rise in generator output voltage; it also increases the spring loading that holds the armature away from the core. Thus, a condition of equilibrium is re-established with the armature in some new position, but with the output voltage still at the required regulated value.

Any reduction of generator speed, within the effective speed range, produces a reduction in generator output voltage thus disturbing regulator armature equilibrium in such a manner that the spring-control force predominates and the armature moves away from the electromagnet core. The carbon pile is re-compressed by this movement to reduce the generator shunt-field circuit resistance and thereby increase generator output voltage, until the regulated output is again brought to a state of equilibrium. When progressive reduction of generator speed results in a condition of maximum pile compression, control of generator output voltage is lost; any further reduction of generator speed, below the lower limit of the effective range, resulting in proportional decrease in output voltage.

When a generator has been run up and connected to its distribution busbar system, the switching on of various requisite consumer services, will impose loads which disturb the equilibrium of the regulator armature. The effect is, in fact, the same as if the generator speed had been reduced, and the regulator automatically takes the appropriate corrective action until the output voltage is stabilized at the critical value. Conversely, a perceptible decrease in load, assuming generator speed to be constant and the regulator armature to be in equilibrium, results in the regulator taking the same action as in the case of an increase in generator speed.

Construction The pile unit is housed within a ceramic tube which, in turn, is enclosed in a solid casing, or more generally, a finned casing for dissipating the heat generated by the pile. The number, diameter, and thickness of the washers which make up the pile, varies according to the specific role of the regulator. Contact at each end of the pile is made by carbon inserts, or in some types of regulator by silver contacts within carbon inserts. The initial pressure of the pile is set by a compression screw acting through the pile on the armature and plate-type control spring

14

which is supported on a bi-metal washer. The washer compensates for temperature effects on voltage coil resistance and on any expansion characteristics of the regulator, thus maintaining constant pile compression. The electromagnet assembly comprises a cylindrical yoke in which is housed the voltage coil, a detachable end-plate and an adjustable soft-iron core. A locking device, usually in the form of screws, is provided to retain the core in a pre-set position.

Depending on the design of generating system, voltage regulators may be of the single-unit type, shown in Fig. 1.15, which operates in conjunction with separate reverse current cut-outs, voltage differential sensing relays and paralleling relays, or integrated with these components to form special control units or panels.

Fig 1.15
Typical single-unit type regulator
1. Armature stop screw
2. Magnet case
3. Heat dissipator
4. Terminal blocks
5. Chassis

Fig. 1.16 shows the circuit arrangement of a typical solid-state voltage regulator as employed with the type of alternator shown in Fig. 1.8. Before going into its operation, however, it will be helpful at this stage, to briefly review the primary function and fundamental characteristics of the device known as the transistor.

The primary function of a transistor is to "transfer resistance" within itself and depending on its connection within a circuit it can turn current "on" and "off" and can increase output signal conditions; in other words, it can act as an automatic switching device or as an amplifier. It has no moving parts and is made up of three regions of a certain material, usually germanium, known as a semiconductor (see also p. 53) and arranged to be in contact with each other in some definite conducting sequence. Some typical transistor contact arrangements are shown in Fig. 1.17 together with the symbols used. The letters "p" and "n" refer to the conductivity characteristic of the germanium and signify positive-type and negative-type respectively. A transistor has three external connections corresponding to the three regions or elements known as the *emitter* which injects the current carriers at one end, the *collector* which collects the current at the other end, and the *base* which controls the amount of current flow. The three elements are arranged to contact each other in sandwich form and in the sequence of either n-p-n or p-n-p. When connected in a circuit the emitter is always forward-biased in order to propel the charged current carriers towards the collector, which is always reverse-biased in order to collect the carriers. Thus, the emitter of an n-p-n transistor has a negative voltage applied to it (with respect to base) so as to repel negative electrons in the forward direction, while a positive voltage is applied to the emitter of a p-n-p transistor so as to repel positively charged "holes" in a forward direction.

Since reverse bias is always applied to collectors then the collector of an n-p-n transistor is made positive with respect to the emitter in order to attract negative electrons. Similarly, the collector of a p-n-p transistor is made negative with respect to the emitter so as to attract positively charged "holes".

The conventional current flow is, of course, opposite to the electron flow and passes through a transistor and the circuit external to it, from emitter to collector and through the base. This is indicated on the symbols adopted for both transistor arrangements, by arrows on the emitter (see Fig. 1.17). Any input voltage that increases the forward bias of the emitter, with respect

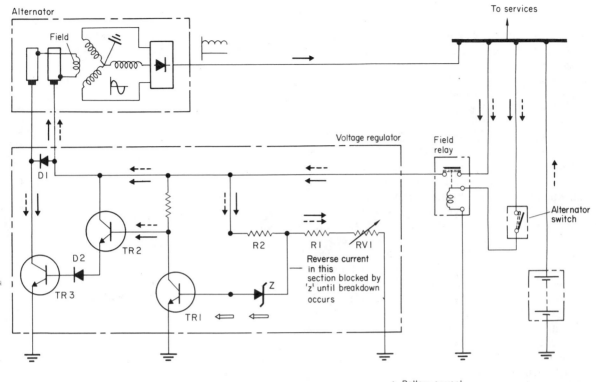

Fig 1.16

Solid-state voltage regulator

- - -► Battery current
——► Rectified current
⊏⊐► Reverse current

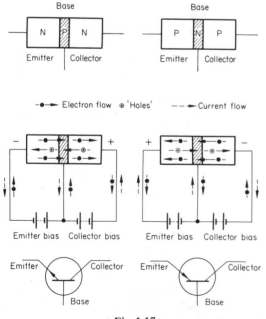

Fig 1.17

Transistor contact arrangements

to the base, increases the emitter-to-collector current flow, and conversely, the current flow is decreased when an input voltage decreases forward bias. The characteristics of transistors are such that small changes in the emitter-base circuit current result in relatively large changes in collector current thereby making transistors efficient amplifying devices. By alternately connecting and disconnecting the base circuit to and from a forward biasing voltage, or similarly, by alternately applying a forward and reverse voltage, base current and thus collector current, can be caused to flow and to cease flowing. In this manner, a transistor can thereby also function as a switching device.

In the regulator circuit shown in Fig. 1.16, the three transistors (TR_1, TR_2 and TR_3) are connected in the n-p-n arrangement. When the system control switch is "on", excitation current flows initially from the battery to the base of TR_2 and through a voltage dividing network made up of resistances R_1, R_2 and RV_1. The purpose of this network in conjunction with the Zener diode "Z" (see also p. 55) is to establish the system-operating voltage. With power applied to the base of TR_2, the transistor is switched on and battery current flows to the collector and emitter junction. The amplified output in the emitter circuit flows to the base of TR_3 thereby switching it on so that the battery current supplied to the field winding can be conducted to ground via the collector-emitter junction of TR_3. When the generator is running, the rotating magnetic field induces an alternating current in the stator and this is rectified and supplied to the d.c. power system of the aircraft.

When the alternator output voltage reaches the pre-set operating value, the current flowing in the reverse direction through the Zener diode causes it to break down and to allow the current to flow to the base of TR_1 thus switching it on. The collector-emitter junction of TR_1 now conducts, thereby diverting current away from the base of TR_2 and switching it off. This action, in turn, switches off TR_3 and so excitation current to the alternator field winding is cut off. The rectifier across the field winding (D_1) provides a path so that field current can fall at a slower rate and thus prevent generation of a high voltage at TR_3 each time it is switched off.

When the alternator output voltage falls to a value which permits the Zener diode to cease conduction, TR_1 will again conduct to restore excitation current to the field winding. This sequence of operation is repeated and the alternator output voltage is thereby maintained at the preset operating value.

Paralleling and Load-Sharing

In multi-engined aircraft, it is generally desirable that the generators driven by each engine should operate in parallel thereby ensuring that in the event of an engine or generator failure, there is no interruption of primary power supply. Parallel operation requires that generators carry equal shares of the system load, and so their output voltages must be as near equal as possible under all operating conditions. As we have already learned, generators are provided with a voltage regulator which exercises independent control over voltage output, but as variations in output and electrical loads can occur, it is essential to provide additional voltage regulation circuits having the function of maintaining balanced outputs and load sharing. The method most commonly adopted for this purpose is that which employs a "load-equalizing circuit" to control generator output via the voltage regulators. The principle as applied to a twin-generator system is illustrated in much-simplified form by Fig. 1.18. The generators are interconnected

on their negative sides, via a series "load-sharing" or "equalizing" loop containing equalizing coils (C_e) each coil forming part of the individual voltage regulator circuits.

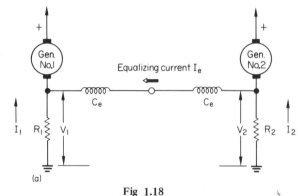

Fig 1.18
Principle of load-sharing

Fig 1.19
Load sharing (carbon pile regulators)

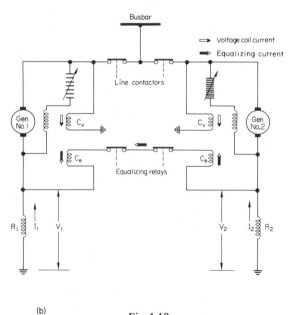

The resistances R_1 and R_2 represent the resistances of the negative sections (interpole windings) of the generators, and under balanced load-sharing conditions the volts drop across each section will be the same,

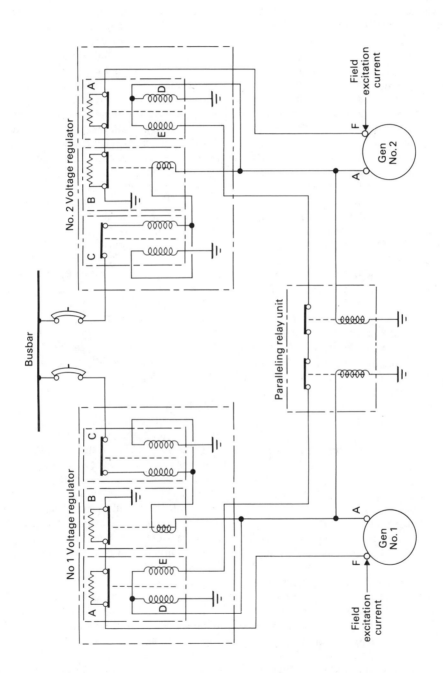

Fig 1.20
Load sharing (vibrating contact regulators)

i.e. $V_1 = I_1 R_1$ and $V_2 = I_2 R_2$. Thus, the net volts drop will be zero and so no current will flow through the equalizing coils.

Let us now assume that generator No. 1 tends to take a somewhat larger share of the total load than generator No. 2. In this condition the volts drop V_1 will now be greater than V_2 and so the negative section of generator No. 1 will be at a lower potential. As a result, a current I_e will flow through the equalizing coils which are connected in such a manner that the effect of I_e is to raise the output voltage of generator No. 2 and reduce that of No. 1, thereby effectively reducing the unbalance in load sharing. Figure 1.19 illustrates the principle as applied to an equalizing circuit which approximates to that of a practical generating system utilizing carbon pile voltage regulators. The equalizing coils are wound on the same magnetic cores as the voltage coils of the regulators, thus, assuming the same unbalanced conditions as before, the current I_e flows in a direction opposite to that flowing through the No. 2 generator voltage regulator coil, but in the same direction as the voltage coil current in No. 1 regulator. The magnetic effect of the No. 2 regulator voltage coil will therefore be weakened resulting in a decrease in carbon pile resistance and an increase in the output of No. 2 generator (see also p. 12), enabling it to take more of the load. The magnetic effect of the No. 1 regulator voltage coil on the other hand, is strengthened, thereby increasing carbon pile resistance and causing No. 1 generator to decrease its output and to shed some of its load. The variations in output of each generator continues until the balanced load-sharing condition is once again restored, whereby the equalizing-circuit loop ceases to carry current.

The principle of paralleling as applied to a twin d.c. generator system utilizing vibrating contact regulators is shown in Fig. 1.20. In this case, the equalizing or paralleling circuit comprises an additional coil "Eq" in the voltage regulation sections "A" of each regulator, and a paralleling relay unit.

When both generators are in operation and supplying the requisite regulated voltage, the contacts in the voltage and current ("B") regulation sections of each regulator are closed. The contacts of the reverse current relays "C" are also closed thereby connecting both generators to the busbar. The outputs from each generator are also supplied to the coils of the paralleling relay unit and so the contacts of its relays are closed. Thus, together with each of the coils "Eq", the equalizing or paralleling circuit is formed

between the generator outputs. Under load-sharing conditions, the current flowing through the coils "Eq" is in the same direction as that through the voltage coils of the voltage regulating sections of each regulator, but in equal and opposite directions at the contacts of the paralleling relay unit.

If the voltage output of one or other generator, e.g. number 1, should rise, there will be a greater voltage input to the voltage regulating section of the number 1 voltage regulator compared to the input at the corresponding section of the number 2 regulator. There will therefore, be an unbalanced flow of current through the equalizing circuit such that the increase in current through the coil "Eq" of the number 1 voltage regulator will now assist the magnetic effect of the voltage coil "D" causing the relay contacts to open. The resistance thereby inserted in the field circuit of number 1 generator reduces its excitation current and its voltage output. Because of the unbalanced condition, the increased current in the equalizing circuit will also flow across the paralleling relay unit contacts to the coil "Eq" in the number 2 voltage regulator so that it opposes the magnetic effect of its associated coil "D".

In paralleled alternator systems using solid-state voltage regulators, any unbalanced condition is detected and adjusted by interconnecting the regulators via two additional paralleling transistors, one in each regulator.

Batteries

In almost all aircraft electrical systems a battery has the following principal functions —

(i) To help maintain the d.c. system voltage under transient conditions. The starting of large d.c. motor-driven accessories, such as inverters and pumps, requires high input current which would lower the busbar voltage momentarily unless the battery was available to assume a share of the load. A similar condition exists should a short circuit develop in a circuit protected by a heavy duty circuit breaker or current limiter. This function possibly applies to a lesser degree on aircraft where the electrical system is predominantly a.c., but the basic principle still holds true.

(ii) To supply power for short term heavy loads when generator or ground power is not available, e.g. internal starting of an engine.

(iii) Under emergency conditions, a battery is intended to supply limited amounts of power. Under these conditions the battery could be the sole remaining source of power to operate essential flight instruments, radio communication equipment, etc., for as long as the capacity of the battery allows.

A battery is a device which converts chemical energy into electrical energy and is made up of a number of cells which, depending on battery utilization, may be of the primary type or secondary type. Both types of cell operate on the same fundamental principle, i.e. the exchange of electrons due to the chemical action of an electrolyte and electrode materials. The essential differences between the two lies in the action that occurs during discharge. In the primary cell this action destroys the active materials of the cell, thus limiting its effective life to a single discharge operation, whereas in the secondary cell the discharge action converts the active material into other forms, from which they can subsequently be electrically reconverted, into the original materials. Thus, a secondary cell can have a life of numerous discharge actions, followed by the action of re-conversion more commonly known as charging. The batteries selected for use in aircraft therefore employ secondary cells and are either of the lead-acid or nickel-cadmium type.

Lead-Acid Secondary Cell

The basic construction of a typical cell is shown in Fig. 1.21. It consists essentially of a positive electrode and a negative electrode, each of which is, in turn, made up of a group of lead-antimony alloy grid plates;

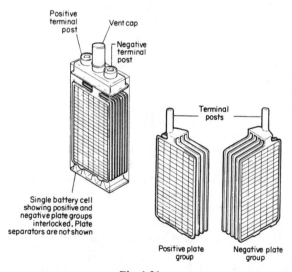

Fig 1.21
Typical lead-acid secondary cell

the spaces of the plates are packed with pastes of active lead materials. The two plate groups are interleaved so that both sides of every positive plate face a negative surface. The plates are prevented from coming into contact with one another by means of separators (not shown) made from materials having high insulating qualities and ability to permit unobstructed circulation of the electrolyte at the plate surfaces. Each group of positive plates and negative plates is connected through a strap to a terminal post at the top and on opposite sides of the cell. The internal resistance of a cell varies immensely with the distance between the positive and negative electrode surfaces; therefore, to obtain the lowest possible resistance the gap between the plates of each group is made as small as is practicable. A cell contains an odd number of plates, the outermost ones belonging to the negative plate group. The reason for this arrangement is that unlike a positive plate a negative plate will not distort when the electromechanical action is restricted to one side only. The plate assemblies of a cell are supported in an acid-proof container.

CHEMICAL ACTION
Each positive plate of a fully-charged cell consists of the lead-antimony alloy grid into which lead peroxide paste (PbO_2) has been forced under pressure. The negative plates are of similar basic structure, but with pure spongy lead (Pb) forced into the grid. The electrolyte consists of two constituents, sulphuric acid (H_2SO_4) and water, which are mixed in such proportions that the relative density is generally about 1·25 to 1·27.

During discharge of the cell, that is, when an external circuit is completed between the positive and negative plates, electrons are transferred through the circuit from lead to lead peroxide and the net result of the chemical reaction is that lead sulphate ($PbSO_4$) forms on both plates. At the same time molecules of water are formed, thus weakening the electrolyte. For all practical purposes, the cell is considered to be discharged when both plates are covered with lead sulphate and the electrolyte has become quite weak.

The cell may be recharged by connecting the positive and negative plates, respectively, to the positive and negative terminals of a d.c. source of slightly higher voltage than the cell. All the foregoing reactions are then reversed; the lead sulphate on the positive plate being restored to lead peroxide, the negative plate restored to spongy lead, and the electrolyte restored to its original relative density.

TYPICAL LEAD-ACID BATTERIES

Two types of lead-acid battery may be found in general use; in one the electrolyte is a free liquid while in the other it is completely absorbed into the plates and separators. An example of the former type of battery is illustrated in Fig. 1.22. The unit has a 24-volts output and consists of two 12-volt cell blocks moulded in high-impact plastic material and housed in an acid-proofed aluminium container. The links interconnecting the cells and cell blocks are sealed and suitably insulated to prevent contact with the container. A plastic tray is fitted on to the top edges of the container and is sealed around the cell vent plugs by rubber pads and plastic sealing rings. The tray forms the base of a chamber for the ventilation of acid vapours. A plastic lid combined with an acid-proofed

aluminium alloy hold-down frame completely encloses the chamber. Connections are provided at each end of the chamber for coupling the pipes from the aircraft's battery compartment ventilation system (see p. 24).

The battery illustrated in Fig. 1.23 utilizes a more specialized form of cell construction than that just described. The plates, active materials and separators are assembled together and are compressed to form a solid block. The active material is an infusorial earth, known as kieselguhr, and is very porous and absorbent. Thus, when the electrolyte is added, instead of remaining free as in the conventional types of battery, it is completely absorbed by the active material. This has a number of advantages; notably improved electro-mechanical activity, no disintegration or shedding of active material, thus preventing internal short-circuits caused by "sludge", low internal resistance and a higher capacity/weight ratio than a conventional battery of comparable capacity.

The cells are assembled as two 12-volt units in monobloc containers made of shock-resistant polystyrene and these are, in turn, housed in a polyester-bonded fibreglass outer container which also supports the main terminal box. A cover of the same material as the case is secured by four bolts on the end flanges of the case.

Nickel-Cadmium Secondary Cell

In this type of cell the positive plates are composed of nickel hydroxide, $Ni(OH)_2$, the negative plates of cadmium hydroxide $Cd(OH)_2$ and the electrolyte is a solution of distilled water and potassium hydroxide (KOH) with a relative density of from 1·24 to 1·30. Batteries made up of these cells have a number of advantages over the lead-acid type, the most notable being their ability to maintain a relatively steady voltage when being discharged at high currents such as during engine starting.

The plates are generally made up by a sintering process and the active materials are impregnated into the plates by chemical deposition. This type of construction allows the maximum amount of active material to be employed in the electrochemical action. After impregnation with the active materials, the plates are stamped out to the requisite size and are built up into positive and negative plate groups, interleaved and connected to terminal posts in a manner somewhat similar to the lead-acid type of cell.

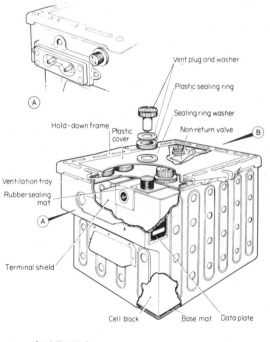

Vent plug and washer
Plastic sealing ring
Sealing ring washer
Non-return valve
Hold-down frame
Plastic cover
Ventilation tray
Rubber sealing mat
Terminal shield
Cell block
Base mat
Data plate

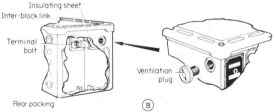

Insulating sheet
Inter-block link
Terminal bolt
Ventilation plug
Rear packing

Fig 1.22
Lead-acid battery (free liquid type)

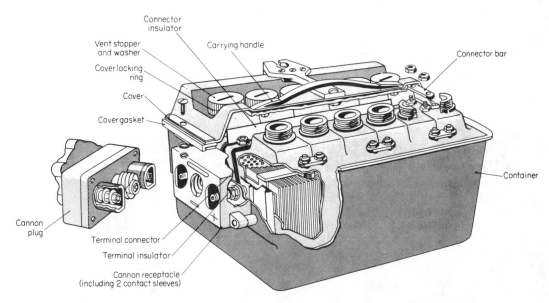

Fig. 1.23
Lead-acid battery (absorbed liquid type)

Insulation is done by means of a fabric-base separator in the form of a continuous strip wound between the plates. The complete plate group is mounted in a sealed plastic container.

CHEMICAL ACTION

During charging, the negative plates lose oxygen and become metallic cadmium. The positive plates are brought to a higher state of oxidation by the charging current until both materials are completely converted; i.e. all the oxygen is driven out of the negative plates and only cadmium remains, the positive plates pick up the oxygen to form nickel oxides. The cell emits gas towards the end of the charging process, and during overcharging; the gas being caused by decomposition of the water component of the electrolyte into hydrogen at the negative plates and oxygen at the positive plates. A slight amount of gassing is necessary to completely charge the cell and so it therefore loses a certain amount of water.

The reverse chemical action takes place during discharging, the negative plates gradually gaining back the oxygen as the positive plates lose it. Due to this interchange there is no gassing on a normal discharge. In this way, the chemical energy of the plates is converted into electrical energy, and the electrolyte is absorbed by the plates to a point where it is not visible from the top of the cell. The electrolyte does not play an active part in the chemical reaction; it is used only to provide a path for current flow.

The chemical reaction of a nickel-cadmium cell is summarized in Table 1.1 and may be compared with that taking place in a lead-acid battery cell.

TYPICAL NICKEL-CADMIUM BATTERY

The construction of a typical battery currently in use is shown in Fig. 1.24. All the cells are linked and contained as a rigid assembly in the case. A space above the cells provides a ventilation chamber which is completely enclosed by a lid held in position by a pair of bolts anchored to the aircraft battery compartment. Acid vapours are drawn out from the chamber via the vents in the battery case and the interconnecting pipes of the aircraft's battery compartment ventilation system.

Capacity of Batteries

The capacity of a battery, or the total amount of energy available, depends upon the size and number of plates. More strictly it is related to the amount of material available for chemical action.

The capacity rating is measured in ampere-hours and is based on the maximum current, in amps, which it will deliver for a known time period, until it is dis-

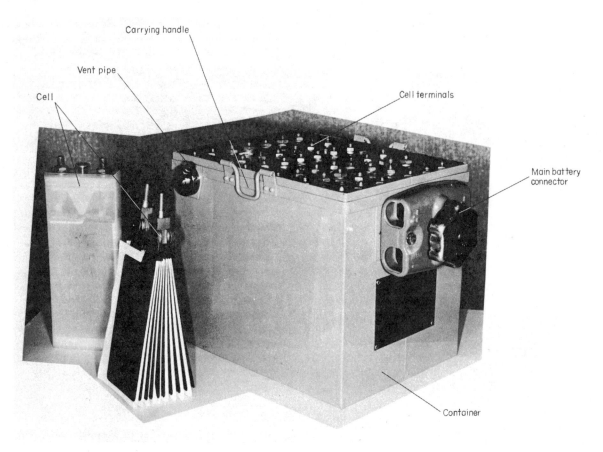

Fig. 1.24
Nickel-cadmium type battery

Table 1.1
Chemical Reactions of Batteries

Battery Type	State of charge	Positive Plate	Negative Plate	Electrolyte
Lead-Acid	charged	PbO_2 (Lead Dioxide)	Pb (Lead)	H_2SO_4 Concentrated Sulphuric Acid
	discharged	$PbSO_4$ (Lead Sulphate)	$Pb SO_4$ (Lead Sulphate)	H_2SO_4 Weak Sulphuric Acid
Nickel-Cadmium	charged	Ni_2O_2 and Ni_2O_3 (Nickel Oxides)	Cd (Cadmium)	KOH (Potassium hydroxide) unaffected by state of charge
	discharged	$Ni(OH)_2$ (Nickel Hydroxide)	$Cd(OH)_2$ (Cadmium Hydroxide)	

charged to a permissible minimum voltage of each cell. The time taken to discharge is called the *discharge rate* and the rated capacity of the battery is the product of this rate and the duration of discharge (in hours). Thus, a battery which discharges 7 A for 5 hours is rated at 35 ampere-hours capacity. Some typical discharge rates of lead-acid and nickel-cadmium batteries are shown in Fig. 1.25.

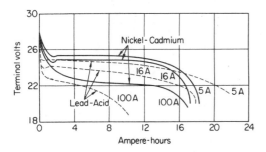

Fig 1.25
Typical discharge rates of lead-acid and nickel-cadmium batteries

STATE OF CHARGE
All batteries display certain indications of their state of charge, and these are of practical help in maintaining operating conditions.

When a lead-acid battery is in the fully-charged condition each cell displays three distinct indications: the terminal voltage reaches its maximum value and remains steady; the relative density of the electrolyte ceases to rise and remains constant; the plates gas freely. The relative density is the sole reliable guide to the electrical condition of the cell of a battery which is neither fully charged nor yet completely discharged. If the relative density is midway between the normal maximum and minimum values then a cell is approximately half discharged.

Checks on the relative density of batteries which do not contain free electrolyte cannot be made; the state of charge being assessed only from voltage indications.

As we have already learned (see p. 21), the electrolyte in the cells of a nickel-cadmium battery does not chemically react with the plates as the electrolyte does in a lead-acid battery. Consequently, the plates do not deteriorate, nor does the relative density of the electrolyte appreciably change. For this reason, it is not possible to determine the state of

charge by checking the relative density. Neither can the charge be determined by a voltage test because of the inherent characteristic that the voltage remains constant over a major part of the discharge cycle. The only possible check that a battery is fully charged is the battery voltage when "on-charge"; additionally, the electrolyte should be at maximum level under these conditions.

Formation of white crystals of potassium carbonate on a properly serviced nickel-cadmium battery installed in an aircraft may indicate that the battery is being overcharged. The crystals form as a result of the reaction of expelled electrolyte vapour with carbon dioxide.

THERMAL RUNAWAY
Batteries are capable of performing to their rated capacities when the temperature conditions and charging rates are within the values specified. In the event that these are exceeded "thermal runaway" can occur, a condition which causes violent gassing, boiling of the electrolyte and finally melting of the plates and casing, with consequent danger to the aircraft structure and jeopardy of the electrical system.

Since batteries have low thermal capacity heat can be dissipated and this results in lowering of the effective internal resistance. Thus, when associated with constant voltage charging, a battery will draw a higher charging current and thereby set up the "runaway" condition of ever-increasing charging currents and temperatures.

In some aircraft, particularly those employing nickel-cadmium batteries, temperature-sensing devices are located within the batteries to provide a warning of high battery temperatures and to prevent overcharging by disconnecting the batteries from the charging source at a predetermined temperature (see also p. 29).

LOCATION OF BATTERIES IN AN AIRCRAFT
Depending on the size of aircraft and on the power requirements for the operation of essential services under emergency conditions, a single battery or several batteries may be provided. When several batteries are employed they are, most often, connected in parallel although in some types of aircraft a series connection is used, e.g. two 14-volt batteries in series, while in others a switching arrangement is

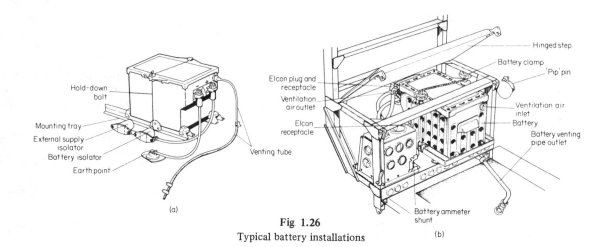

Fig. 1.26
Typical battery installations

incorporated for changing from one method of connection to the other.

Batteries are installed in individual compartments specially designed and located to provide adequate heat dissipation, ventilation of gases and protection of airframe structure against corrosive elements. At the same time batteries should be located as near to the main and battery busbars as physically possible in order to avoid the use of long leads and consequent high resistance. Batteries are normally mounted on, and clamped to, a tray secured to the aircraft structure. The tray forms a catchment for any acid which may escape from the battery. Trays may be of any material which is acid-proof, non-absorbent and resistant to reasonable impacts. Many reinforced plastics are suitable but metal trays are, on the whole, undesirable. Where metal trays are unavoidable they are treated with an anti-corrosive paint or, in some cases, sprayed or coated with p.v.c. The structure under and around the battery area is also treated to avoid corrosive attack by acid fumes and spray. Batteries are securely clamped and anchored to their structure to prevent their being torn loose in the event of a crash landing, thus minimizing the risk of fire. Two typical battery installations are illustrated in Fig. 1.26.

Venting of batteries and battery compartments may take various forms since it depends largely on the installation required for a particular type of aircraft. Rubber or other non-corrosive pipes are usually employed as vent lines which terminate at ports in the fuselage skin so that the airflow over it draws air through the pipes by a venturi action. In some cases, acid traps, in the form of polythene bottles, are

inserted in the lines to prevent acid spray being ejected on to the outer-skin of the aircraft.

In the installation shown in Fig. 1.26(b) fumes and gases generated by the battery are extracted by the difference of pressure existing across the aircraft. During normal flight air tapped from the cabin pressurization system enters the battery ventilation chamber and continues through to the outside of the aircraft. On the ground, when no pressure differential exists, a non-return valve fitted in the air inlet prevents fumes and gases from escaping into the aircraft. These typical venting arrangements are illustrated schematically in Fig. 1.27.

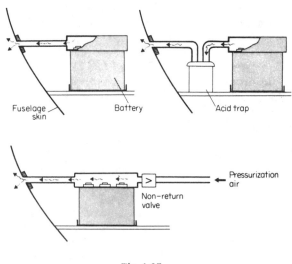

Fig. 1.27
Battery venting arrangements

BATTERY CONNECTIONS

The method of connecting batteries to their respective busbars or power distribution points, depends largely on the type of battery employed, and on the aircraft's electrical system. In some cases, usually on the smaller types of aircraft, the connecting leads are provided with forked lugs which fit on to the appropriate battery terminals. However, the method most commonly employed is the plug and socket type connector shown in Fig. 1.28. It provides better connection and, furthermore, shields the battery terminals and cable terminations.

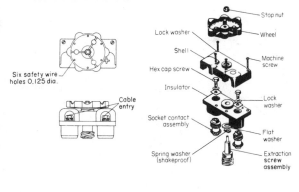

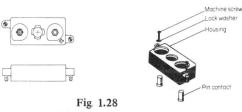

Fig. 1.28
Battery plug connector

The socket comprises a plastic housing, incorporated as an integral part of the battery, two shrouded plug pins and the female threaded portion of a quick-start thread lead-screw. The plug consists of a plastic housing incorporating two shrouded spring-loaded sockets and terminals for the connection of battery leads, and the male half of the mating lead-screw operated by a handwheel. The two halves, on being engaged, are pulled into position by the lead-screw which thereafter acts as a lock. Reverse rotation of the handwheel separates the connector smoothly with very little effort. In this way high contact pressures and low resistance connections are possible and are consistently maintained.

TYPICAL BATTERY SYSTEMS

Figure 1.29 shows the circuit arrangement for a battery system which is employed in a current type of turboprop airliner; the circuit serves as a general guide to the methods adopted. Four batteries, in parallel are directly connected to a battery busbar which, in the event of an emergency, supplies power for a limited period to essential consumer services, i.e. radio, fire-warning and extinguishing systems, a compass system, etc. Direct connections are made to ensure that battery power is available at the busbar at all times.

The batteries also require to be connected to ensure that they are maintained in a charged condition. In the example illustrated this is accomplished by connecting the batteries to the main d.c. busbar via a battery relay, power selector switch and a reverse current circuit breaker.

Under normal operating conditions of the d.c. supply system, the power selector switch is set to the "battery" position (in some aircraft this may be termed the "flight" position) and, as will be noted, current flows from the batteries through the coil of the battery relay, the switch, and then to ground via the reverse current circuit breaker contacts. The current flow through the relay coil energizes it, causing the contacts to close thereby connecting the batteries to the main busbar via the coil and second set of contacts of the reverse current circuit breaker. The d.c. services connected to the main busbar are supplied by the generators and so the batteries will also be supplied with charging current from this source.

Under emergency conditions, e.g. a failure of the generator supply or main busbar occurs, the batteries must be isolated from the main busbar since their total capacity is not sufficient to keep all services in operation. The power selector switch must therefore be put to the "off" position, thus de-energizing the battery relay. The batteries then supply the essential services for the time period pre-calculated on the basis of battery capacity and current consumption of the essential services.

The reverse current circuit breaker in the system shown is of the electromagnetic type and its purpose is to protect the batteries against heavy current flow from the main busbar. Should this happen the current reverses the magnetic field causing the normally closed contacts to open and thereby interrupt the circuit between the batteries and main busbar, and the battery relay coil circuit.

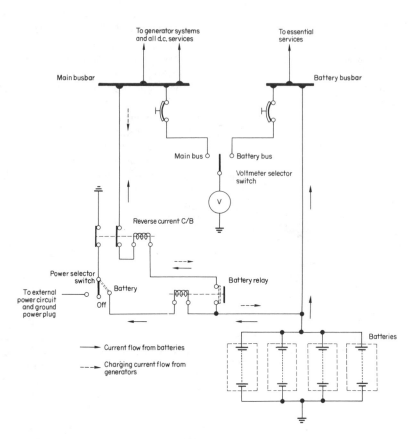

Fig 1.29
Typical battery system circuit

The battery system in some types of turboprop powered aircraft is so designed that the batteries may be switched from a parallel configuration to a series configuration for the purpose of starting an engine from the batteries. The circuit arrangement of one such system using two 24-volt nickel-cadmium batteries is shown in simplified form in Fig. 1.30.

Under normal parallel operating conditions, battery 1 is connected to the battery busbar via its own battery relay, and also contacts 1a-1b of a battery switching relay. Battery 2 is directly connected to the busbar via its relay.

When it is necessary to use the batteries for starting an engine, i.e. to make an "internal" start, both batteries are first connected to the battery busbar in the normal way, and the 24-volt supply is fed to the starter circuit switch from the busbar. Closing of the starter switch energizes the corresponding starter relay, and at the same time the 24-volt supply is fed via the starting circuit, to the coil of the battery switching relay thereby energizing it. Contacts 1a-1b of the relay are now opened to interrupt the direct connection between battery 1 and the busbar. Contacts 3a-3b are also opened to interrupt the grounded side of battery 2. However, since contacts 2a-2b of the switching relay are simultaneously moved to the closed position, they connect both batteries in series so that 48 volts is supplied to the busbar and to the starter motor.

After the engine has started and reached self-sustaining speed, the starter relay automatically de-energizes and the battery switching relay coil circuit is interrupted to return the batteries to their normal parallel circuit configuration.

The power selector switches are left in the "battery" position so that when the engine-driven generator is switched onto the busbar, charging current can flow to the batteries.

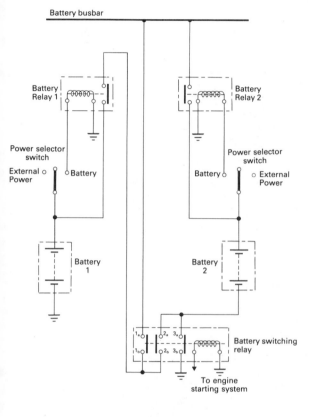

Fig 1.30
Parallel/series connection of batteries

of charging it. The purpose of the fuse in the closing circuit is to interrupt the charge in the event of a "shorted" battery.

When the battery is being charged in this manner, the voltage and current output from the external power unit must be properly regulated.

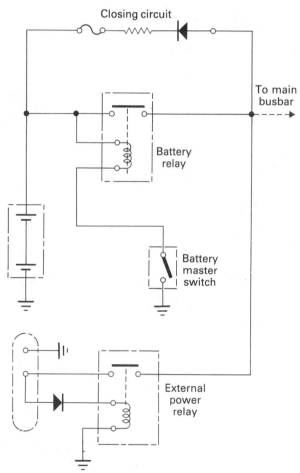

Fig 1.31
Battery charging from external power

Battery Charging from External Power

In some single-engined aircraft systems, the battery may be charged when an external power unit is plugged into the aircraft. This is achieved by a battery relay closing circuit connected across the main contacts of the relay as shown in Fig. 1.31.

With the external power connected and switched on, power is available to the battery relay output terminal via the closed contacts of the external power relay. At the same time, power is applied to the battery relay closing circuit via its diode and resistor which reduces the voltage to the input side of the battery relay's main contacts and coil.

When the battery master switch is selected to "on", sufficient current flows through the coil of the battery relay to energize it. The closed contacts of the relay then allow full voltage from the external power unit to flow to the battery for the purpose

"On-board" Battery Charger Units

In most types of turbojet transport aircraft currently in service, the battery system incorporates a separate unit for maintaining the batteries in a stage of charge. Temperature-sensing elements are also normally provided in order to automatically isolate the charging

28

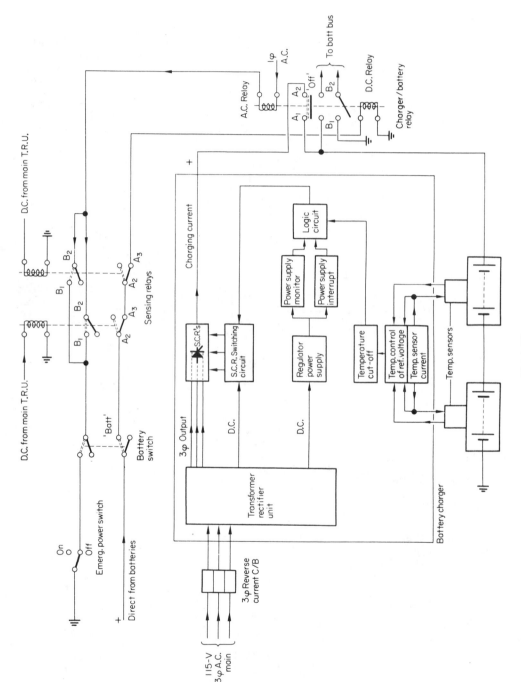

Fig 1.32

In-situ battery charging system

circuit whenever there is a tendency for battery over-heating to occur. The circuits of "on-board" charger units as they are generally termed, vary between aircraft types, and space limits description of all of them. We may however, consider two examples which highlight some of the variations to be found.

The much simplified circuit shown in Fig. 1.32 is based on the system adopted for the McDonnell Douglas DC-10. In this particular application, the required output of 28 volts is achieved by connecting two 14-volt batteries in series. Furthermore, and unlike the system shown in Fig. 1.29, the batteries are only connected to the battery busbar whenever the normal d.c. supply (in this case from transformer/rectifier units) is not available. Connection to the busbar and to the charger unit is done automatically by means of a "charger/battery" relay and by sensing relays.

When power is available from the main generating system, d.c. is supplied to the battery busbar from a transformer/rectifier unit and, at the same time, to the coils of the sensing relays. With the relays ener-gized, the circuit through contacts A2-A3 is inter-rupted while the circuits through contacts B1-B2 are made. The battery switch, which controls the operation of the charger/battery relay, is closed to the "batt" position when the main electrical power is available, and the emergency power switch is closed in the "off" position.

The charger/battery relay is of the dual type, one relay being a.c. operated and the other d.c. operated. The a.c. relay coil is supplied with power from one phase of the main three-phase supply to the battery charger, and as will be noted from the diagram, the relay is energized by current passing to ground via the contacts B1-B2 of the sensing relays, the battery switch and the emergency switch. Energizing of the relay closes the upper set of contacts (A1-A2) to connect the d.c. positive output from the battery charger to the batteries, thereby supplying them with charging current.

In the event of main power failure, the battery charger will become inoperative, the a.c. charger relay will de-energize to the centre off position, and the two sensing relays will also de-energize, thereby opening the contacts B1-B2 and closing the contacts A2-A3. The closing of contacts A2-A3 now permits a positive supply to flow direct from the battery to the coil of the d.c. battery relay, which on being energized also actuates the a.c. relay, thereby closing contacts B1-B2 which connect the batteries direct to the battery

busbar. The function of the battery relay contacts is to connect a supply from the battery busbar to the relays of an emergency warning light circuit. The charg-ing unit converts the main three-phase supply of 115/200 volts a.c. into a controlled d.c. output at constant current and voltage, via a transformer and a full-wave rectifying bridge circuit made up of silicon rectifiers and silicon controlled rectifiers (see also p. 57). The charging current is limited to approximately 65 A, and in order to monitor this and the output voltage as a function of battery temperature and voltage, temper-ature-sensing elements within the batteries are connected to the S.C.R. "gates" via a temperature and reference voltage control circuit, and a logic circuit. Thus, any tendency for overcharging and overheating to occur is checked by such a value of gate circuit current as will cause the S.C.R. to switch off the charging current supply.

The second example shown in Figs. 1.33 and 1.34, is based on that used in the Boeing 737. The charger operates on 115 volt 3-phase a.c. power supplied from a "ground service" busbar, which in turn, is normally powered from the number 1 generator busbar, and/or from an external power source (see page 27). Thus, the aircraft's battery is maintained in a state of charge both in flight and on the ground.

In flight the a.c. supply is routed to the charger through the relaxed contacts of a battery charger transfer relay and an APU start interlock relay. The d.c. supply for battery charging is obtained from a transformer-rectifier unit within the charger, and it maintains cell voltage levels in two modes of operation: high and low. Under normal operating conditions of the aircraft's power generation system, the charging level is in the high mode since as will be noted from the diagram, the mode control relay within the charger is energized by a rectified output through the battery thermal switch, and the relaxed contacts of both the battery bus relay and the external power select relay. Above 16 amps the charger acts as an unregulated transformer-rectifier unit, and when the battery has sufficient charge that the current tends to go below 16 amps, the charging current is abruptly reduced to zero. The current remains at zero until the battery voltage drops below the charge voltage, at which time the charger provides the battery with a pulsed charge and the process is repeated. The pulsing continues until the control circuits within the charger change the operation to the low mode, approximately two

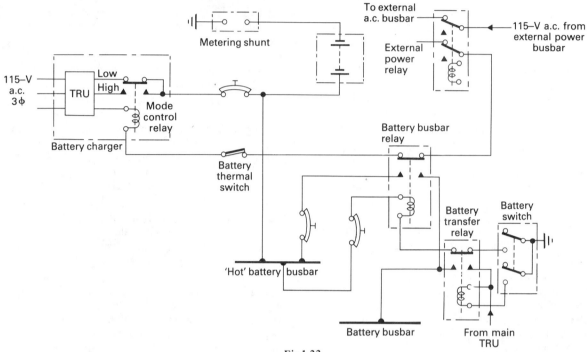

Fig 1.33
Battery charger control circuit

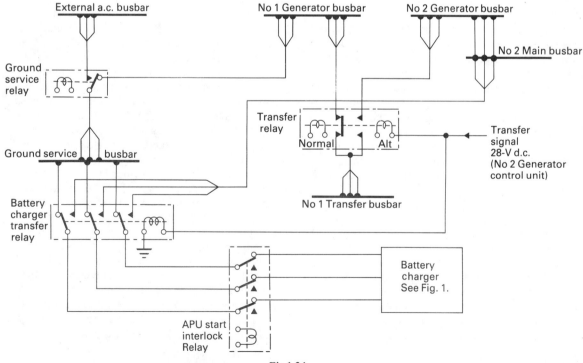

Fig 1.34
Battery charger a.c. input

minutes after pulse charging commences.

In the event that the number one generator supply fails there will be a loss of a.c. power to the ground service busbar, and therefore, to the battery charger. However, with number two generator still on line, a transfer signal from its control unit is automatically supplied to the coil of the battery charger transfer relay, and as may be seen from Fig. 1.34, its contacts change over to connect the charger to the a.c. supply from number two generator, and so charger operation is not interrupted.

The APU start interlock relay is connected in parallel with a relay in the starting circuit of the APU, and is only energized during the initial stage of starting the APU engine. This prevents the starter motor from drawing part of its heavy starting current through the battery charger. The interlock relay releases automatically when the APU engine reaches 35% rev/min.

In addition to the control relay within the battery charger, there are three other ways in which the charging mode can be controlled, each of them fulfilling a protective role by interrupting the ground circuit to the mode control relay and so establishing a low mode of charge. They are — (i) opening of the battery thermal switch in the event of the battery temperature exceeding 46 °C (115 °F); (ii) loss of d.c. power from the designated transformer-rectifier unit causing the battery transfer relay to relax and the battery bus relay to energize; and (iii) energizing of the fuelling panel power select relay when external a.c. power is connected to the aircraft. The latter is of importance since if the charger was left to operate in the high mode, then any fault in the regulation of the external power supply could result in damage to the aircraft battery.

CHAPTER TWO

Alternating Current Power Supplies

FUNDAMENTAL PRINCIPLES

Before studying the operation of some typical generating systems currently in use it will be of value to recapitulate certain of the fundamentals of alternating current behaviour, and of terminology commonly used.

Cycle and Frequency

The voltage and current produced by the generator of an a.c. system build up from zero to a maximum of one polarity, then decay to zero, build up to a maximum of opposite polarity, and again decay to zero. This sequence of build up and reversal follows a sine wave form and is called a *cycle* and the number of cycles in unit time (usually one second) is called the *frequency* (see Fig. 2.1). The unit of frequency measurement is the Hertz (Hz).

In a conventional generator, the frequency is

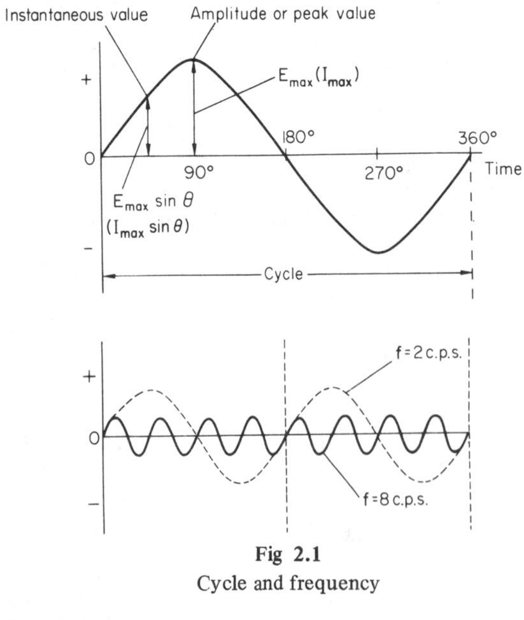

Fig 2.1
Cycle and frequency

dependent upon the speed of rotor rotation within its stator and the number of poles. Two poles of a rotor must pass a given point on the stator every cycle; therefore:

$$\text{Frequency (c.p.s.)} = \frac{\text{rev/min} \times \text{ pairs of poles}}{60}$$

For example, with a 6-pole generator operating at 8000 r.p.m.,

$$\text{Frequency} = \frac{8000 \times 3}{60} = 400 \text{ c.p.s. or } 400 \text{ Hz}$$

For aircraft constant frequency systems (see p. 46) 400 Hz has been adopted as the standard.

Instantaneous and Amplitude Values

At any given instant of time the actual value of an alternating quantity may be anything from zero to a maximum in either a positive or negative direction; such a value is called an *Instantaneous Value*. The *Amplitude* or *Peak Value* is the maximum instantaneous value of an alternating quantity in the positive and negative directions.

The wave form of an alternating e.m.f. induced in a single-turn coil, rotated at a constant velocity in a uniform magnetic field, is such that at any given point in the cycle the instantaneous value of e.m.f. bears a definite mathematical relationship to the amplitude value. Thus, when one side of the coil turns through $\theta°$ from the zero e.m.f. position and in the positive direction, the instantaneous value of e.m.f. is the product of the amplitude (E_{max}) and the sine of θ or, in symbols:

$E_{inst} = E_{max} \sin \theta$

Similarly, the instantaneous value of current is $I_{inst} = I_{max} \sin \theta$.

Root Mean Square Value

The calculation of power, energy, etc., in an a.c. circuit is not so perfectly straightforward as it is in a d.c. circuit because the values of current and voltage are changing throughout the cycle. For this reason, therefore, an arbitrary "effective" value is essential. This value is generally termed the Root Mean Square (r.m.s.) value (see Fig. 2.2). It is obtained by taking a

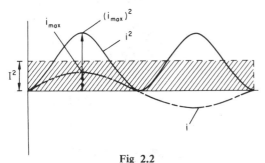

Fig 2.2
R.M.S. value of alternating current

number of instantaneous values of voltage or current, whichever is required, during a half cycle, squaring the values and taking their mean value and then taking the square root. Thus, if six values of current "I" are taken, the mean square value is:

$$\frac{I_1{}^2 + I_2{}^2 + I_3{}^2 + I_4{}^2 + I_5{}^2 + I_6{}^2}{6}$$

and the r.m.s. value is:

$$\sqrt{\frac{I_1{}^2 + I_2{}^2 + I_3{}^2 + I_4{}^2 + I_5{}^2 + I_6{}^2}{6}}$$

The r.m.s. value of an alternating current is related to the amplitude or peak value according to the wave form of the current. For a sine wave the relationship is given by:

$$\text{r.m.s.} = \frac{\text{Peak}}{\sqrt{2}} = 0.707 \text{ Peak}$$

$$\text{Peak} = \sqrt{2} \text{ r.m.s.} = 1.414 \text{ r.m.s.}$$

Phasing and Phase Relationships

In connection with a.c. generating systems and associated circuits, the term "phase" is used to indicate the number of alternating currents being produced and/or carried simultaneously by the same circuit. Furthermore, it is used in designating the type of generating system and/or circuit, e.g. a "single-phase" system or one producing single-phase current, and a "polyphase" system or one producing several single

alternating currents differing in phase. Aircraft poly-phase systems and circuits are normally three-phase, the three currents differing in phase from each other by 120 electrical degrees.

The current and voltage in an a.c. circuit have the same frequency, and the wave form of the alternating quantities is similar, i.e. if the voltage is sinusoidal then the current is also sinusoidal. In some circuits the flow of current is affected solely by the applied voltage so that both voltage and current pass through zero and attain their peaks in the same direction simultaneously; under these conditions they are said to be "in phase". In many circuits, however, the current flow is influenced by magnetic and electrostatic effects set up in and around the circuit, and although at the same frequency, voltage and current do not pass through zero at the same instant. In these circumstances the voltage and current are said to be "out of phase", the difference between corresponding points on the wave-forms being known as the phase difference. The term "phase angle" is quite often used, and is synonymous with phase difference when expressed in angular measure. The phase relationships for the three basic forms of a.c. circuits, namely, pure resistive, inductive and capacitive, are illustrated in Fig. 2.3.

In a pure resistive circuit (Fig. 2.3(a)) the resistance is constant, therefore magnetic and electrostatic effects are absent, and the applied voltage is the only factor affecting current flow. Thus, voltage and current are "in phase" in a resistive circuit.

In a pure inductive circuit (normally some resistance is always present) voltage and current are always out of phase. This is due to the fact that a magnetic field surrounds the conductors, and since it too continually changes in magnitude and direction with the alternating current, a self-induced or "reactance" e.m.f. is set up in the circuit, to oppose the change of current in the circuit. As a result the rise and fall of the current is delayed and as may be seen from Fig. 2.3(b) the current "lags" the voltage by 90 degrees.

Capacitance in an a.c. circuit also opposes the current flow and causes a phase difference between applied voltage and current but, as may be noted from Fig. 2.3(c), the effect is the reverse to that of inductance, i.e. the current "leads" the voltage by 90 degrees.

Where the applied voltage and current are out of phase by 90 degrees they are said to be in *quadrature*.

A three-phase circuit is one in which three voltages are produced by a generator with three coils so spaced within the stator, that the three voltages generated are equal but reach their amplitude values at different

34

times. For example, in each phase of a 400 Hz, three-phase generator, a cycle is generated every 1/400 second. In its rotation, a magnetic pole of the rotor passes one coil and generates a maximum voltage; one-third of a cycle (1/1200 second) later, this same pole passes another coil and generates a maximum voltage in it. Thus, the amplitude values generated in the three coils are always one-third of a cycle (120 electrical degrees; 1/1200 second) apart.

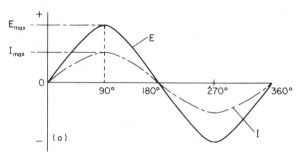

(a)

Pure resistive — in phase

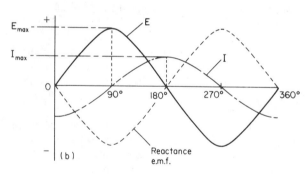

(b)

Pure inductive — I lags behind E

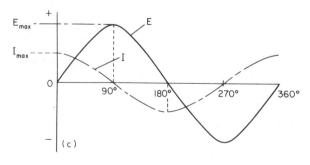

(c)

Pure capacitive — I leads E

Fig 2.3
A.C. circuits phase relationship

The interconnection of the coils to form the three phases of a basic generator, and the phase sequence, is shown in Fig. 2.4. The output terminals of generators are marked to show the phase sequence, and these terminals are connected to busbars which are identified correspondingly.

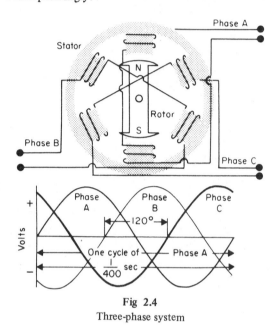

Fig 2.4
Three-phase system

Interconnection of Phases

Each phase of a three-phase generator may be brought out to separate terminals and used to supply separate groups of consumer services. This, however, is an arrangement rarely encountered in practice since pairs of "line" wires would be required for each phase and would involve uneconomic use of cable. The phases are, therefore, interconnected normally by either of the two methods shown in Fig. 2.5.

The "Star" connection ((a)) is commonly used in generators. One end of each phase winding is connected to a common point known as the *neutral* point, while the opposite ends of the windings are connected to three separate lines. Thus, two-phase windings are connected between each pair of lines. Since similar ends of the windings are joined, the two phase e.m.f.s are in opposition and out of phase and the voltage between lines (E_L) is the phase voltage (E_{ph}) multiplied by $\sqrt{3}$. For example, if E_{ph} is 120 volts, then E_L equals 120 x 1·732, or 208 volts approx. As far as line and phase currents are concerned, these are equal to each other in this type of circuit connection.

If necessary, consumer services requiring only a single-phase supply can be tapped into a three-phase star-connected system with a choice of two different voltage levels. Thus, by connecting from one phase to neutral or ground, we obtain a single-phase 120 volts supply while connecting across any pair of lines we can obtain a single-phase 208 volts supply.

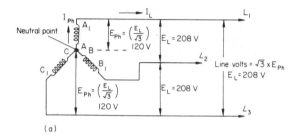

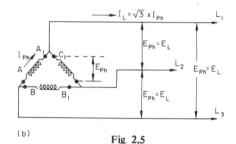

Fig. 2.5
Interconnection of phases
(a) "Star" connection
(b) "Delta" connection

Figure 2.5(b) illustrates the "Delta" method of connection, the windings being connected in series to form a closed "mesh" and the lines being connected to the junction points. As only one phase winding is connected between each pair of lines then, in the delta method, line voltage (E_L) is always equal to phase voltage (E_{ph}). The line current, however, is the difference between the phase currents connected to the line and is equal to the phase current (I_{ph}) multiplied by $\sqrt{3}$.

Generator Power Ratings

The power ratings of a.c. generators are generally given in kilovolt-amperes (kVA) rather than kilowatts (kW) as in the case of d.c. machines. The primary reason for this is due to the fact that in calculating the power, account must be taken of the difference between the true or effective power, and the apparent power. Such a difference arises from the type of

circuit which the generator is to supply and the phase relationships of voltage and current, and is expressed as a ratio termed the power factor (P.F.). This may be written:

$$P.F. = \frac{\text{Effective Power (kW)}}{\text{Apparent Power (kVA)}}$$

$$= \text{cosine phase angle } \varphi$$

If the voltage and current are in phase (as in a resistive circuit) the power factor is 100 per cent or unity, because the effective power and apparent power are equal; thus, a generator rated at 100 kVA in a circuit with a P.F. of unity will have an output 100 per cent efficient and exactly equal to 100 kW.

When a circuit contains inductance or capacitance, then as we have already seen (p. 33) current and voltage are not in phase so that the P.F. is less than unity. The vector diagram for a current I lagging a voltage E by an angle φ is shown in Fig. 2.6. The current is resolved into two components at right angles, one in phase with E and given by I cos φ, and the other in quadrature and given by I sin φ. The in-phase component is called the active, wattful or working component (kW) and the quadrature component is the idle, wattless or reactive component (kVAR). The importance of these components will be more apparent when, later in this chapter, methods

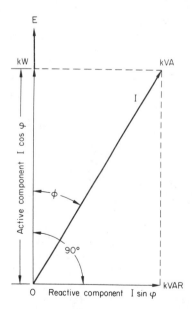

Fig 2.6
Components of current due to phase difference

of load sharing between generators are discussed.

Most a.c. generators are designed to take a proportion of the reactive component of current through their windings and some indication of this may be obtained from the information given on the generator data plate. For example, the output rating may be specified as 40 kVA at 0·8 P.F. This means that the maximum output in kW is 0·8 x 40 or 32 kW, but that the product of volts and amperes under all conditions of P.F. must not exceed 40 kVA.

FREQUENCY-WILD SYSTEMS

A frequency-wild system is one in which the frequency of its generator voltage output is permitted to vary with the rotational speed of the generator. Although such frequency variations are not suitable for the direct operation of all types of a.c. consumer equipment, the output can (after constant voltage regulation) be applied directly to resistive load circuits such as electrical de-icing systems, for the reason that resistance to alternating current remains substantially constant, and is independent of frequency.

Generator Construction

The construction of a typical frequency-wild generator utilized for the supply of heating current to a turbo-propeller engine de-icing system is illustrated in Fig. 2.7. It has a three-phase output of 22 kVA at 208 volts and it supplies full load at this voltage through a frequency range of 280 to 400 Hz. Below 280 Hz the field current is limited and the output relatively reduced. The generator consists of two major assemblies: a fixed stator assembly in which the current is induced, and a rotating assembly referred to as the rotor. The stator assembly is made up of high permeability laminations and is clamped in a main housing by an end frame having an integral flange for mounting the generator at the corresponding drive outlet of an engine-driven accessory gear-box. The stator winding is star connected, the star or neutral point being made by linking three ends of the winding and connecting it to ground (see also p. 35). The other three ends of the winding are brought out to a three-way output terminal box mounted on the end frame of the generator. Three small current transformers are fitted into the terminal box and form part of a protection system known as a Merz-Price system (see p. 122).

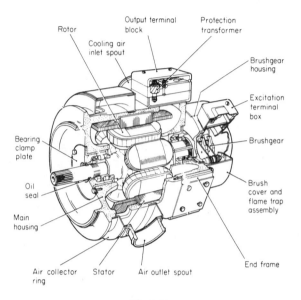

Fig 2.7
Frequency-wild generator

The rotor assembly has six salient poles of laminated construction; their series-connected field windings terminate at two slip rings secured at one end of the rotor shaft. Three spring-loaded brushes are equispaced on each slip ring and are contained within a brush-gear housing which also forms a bearing support for the rotor. The brushes are electrically connected to d.c. input terminals housed in an excitation terminal box mounted above the brush-gear housing. The terminal box also houses capacitors which are connected between the terminals and frame to suppress interference in the reception of radio signals. At the drive end, the rotor shaft is serrated and an oil seal, housed in a carrier plate bolted to the main housing, is fitted over the shaft to prevent the entry of oil from the driving source into the main housing.

The generator is cooled by ram air (see also Chapter 1, p. 9) passing into the main housing via an inlet spout at the slip ring end, the air escaping from the main housing through ventilation slots at the drive-end. An air-collector ring encloses the slots and is connected to a vent through which the cooling air is finally discharged. Provision is made for the installation of a thermally-operated switch to cater for an overheat warning requirement.

CONSTANT FREQUENCY SYSTEMS

In the development of electrical power supply systems, notably for large aircraft, the idea was conceived of an "all a.c." system, i.e. a primary generating system to meet all a.c. supply requirements, in particular those of numerous consumer services dependent on constant-frequency, to allow for paralleled generator operation, and to meet d.c. supply requirements via transformer and rectifier systems.

A constant frequency is inherent in an a.c. system only if the generator is driven at a constant speed. The engines cannot be relied upon to do this directly and, as we have already learned, if a generator is connected directly to the accessory drive of an engine the output frequency will vary with engine speed. Some form of conversion equipment is therefore required and the type most widely adopted utilizes a transmission device interposed between the engine and generator, and which incorporates a variable-ratio drive mechanism. Such a mechanism is referred to as a constant-speed drive (CSD) unit and an example is shown in Fig. 2.8.

The unit employs a hydromechanical variable-ratio drive which in its basic form (see Fig. 2.9)

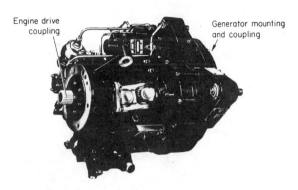

Fig 2.8
Constant speed drive unit

consists of a variable displacement hydraulic unit, a fixed displacement hydraulic unit and a differential gear. The power used to drive the generator is controlled and transmitted through the combined effects of the three units, the internal arrangement of which are shown in Fig. 2.10. Oil for system operation is supplied from a reservoir via charge pumps within the unit, and a governor.

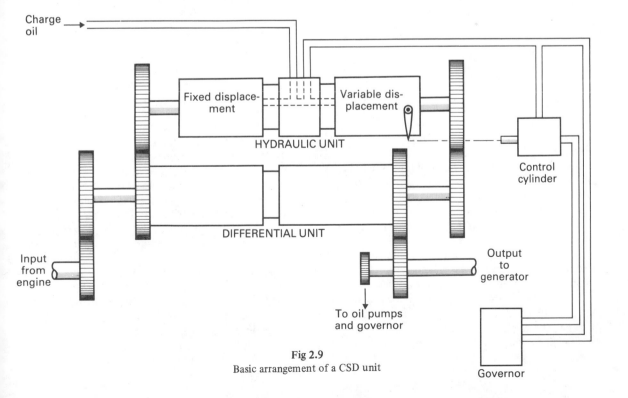

Fig 2.9
Basic arrangement of a CSD unit

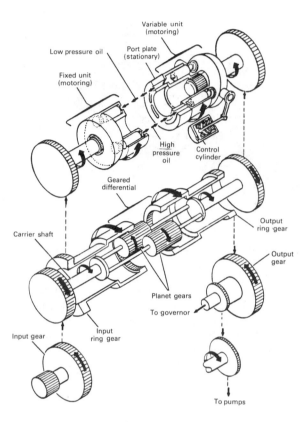

Low pressure oil

Variable unit (motoring)

Port plate (stationary)

Fixed unit (motoring)

High pressure oil

Control cylinder

Geared differential

Carrier shaft

Output ring gear

Output gear

Planet gears

To governor

Input gear

Input ring gear

To pumps

Fig 2.10
Underdrive phase

The variable displacement unit consists of a cylinder block, reciprocating pistons and a variable angle wobble or swash plate, the latter being connected to the piston of a control cylinder. Oil to this cylinder is supplied from the governor. The unit is driven directly by the input gear and the differential planet gear carrier shaft, so that its cylinder block always rotates (relative to the port plate and wobble plate) at a speed proportional to input gear speed and always in the same direction. When the control cylinder moves the wobble plate to some angular position, the pistons within the cylinder block are moved in and out as the block rotates, and so the charge oil is compressed to a high pressure and then "ported" to the fixed displacement unit. Thus, under these conditions the variable displacement unit functions as a hydraulic pump.

The supply of charge oil to the unit's control valve is controlled by a governor valve which is spring biased, flyweight operated and driven by the output gear driving the generator. It therefore responds to changes in transmission output speed.

The fixed displacement unit is similar to the variable displacement unit, except that its wobble plate which has an inclined face, is fixed and has no connection with the control cylinder. When oil is pumped to the fixed displacement unit by the variable one, it functions as a hydraulic motor and its direction of rotation and speed is determined by the volume of oil pumped to it. It can also function as a pump and therefore supply the variable displacement hydraulic unit.

The differential gear consists of a carrier shaft carrying two meshing (1:1 ratio) planet gears, and a gear at each end; one meshing with the input gear and the other with the gear which continuously drives the variable displacement unit cylinder block. The carrier shaft always rotates in the same direction and at a speed which, via the input gear, varies with engine speed. Surrounding the carrier shaft are two separate "housings", and since they have internal ring gears meshing with the planet gears, then they can be rotated differentially. Each housing also has an external ring gear; one (input ring gear) meshing with the fixed displacement unit gear, and the other (output ring gear) meshing with the output gear drive to the generator. Thus, with the CSD in operation, the output ring gear "housing" serves as the continuous drive transmission link between engine and generator. Since the input ring gear "housing" is geared to the fixed displacement hydraulic unit, then depending on whether this unit is acting as a motor or a pump, the "housing" can rotate in the same direction as, or opposite to, that of the carrier shaft and the output ring gear "housing". In this way, speed is added to, or subtracted from, the engine speed, and through the gear ratio (2:1) between the ring gears and the carrier shaft planet gears, the output ring gear "housing" rotational speed will be appropriately adjusted to maintain constant governor speed.

When the input speed, via the input gear, is sufficient to produce the required output speed, the drive to the generator is transmitted straight through the differential and output ring gear. The variable displacement hydraulic unit cylinder block is continuously rotating, but the position of its wobble plate is such that no charge oil is pumped to the fixed displacement unit. The cylinder block of this unit and the output ring gear "housing" do not, therefore, rotate during straight through drive.

If the input speed supplied to the transmission exceeds that needed to produce the required output speed, the governor in sensing the speed difference will cause oil to flow away from the control valve. In this condition, the transmission is said to be operating in the *underdrive* phase and is shown in Fig. 2.10. The control valve changes the angular position of the variable displacement unit's wobble plate so that the volume of oil for accommodating the oil in the bores of the cylinder block is increased, allowing oil to be pumped at high pressure from the fixed displacement unit. The pressure of the pistons against the inclined face of the unit now causes its cylinder block to rotate in the same direction as that of the variable displacement unit. This rotation is transmitted to the input ring gear "housing" of the differential unit, so that it will rotate in the same direction as the output ring gear "housing", and the carrier shaft. Because the input ring gear "housing" is now rotating in the same direction as the carrier shaft then the speed of the freely rotating planet gear meshing with the housing will be reduced. The speed of the second planet gear will also be reduced in direct ratio thereby reducing the speed of the output ring gear "housing". This hydromechanical process of speed subtraction continues until the required generator drive speed is attained at which the transmission will revert to straight-through drive operation.

When the input speed supplied to the transmission is lower than that needed to produce the required output speed, the governor causes charge oil to be supplied to the control valve. In this condition, the transmission is said to be operating in the *overdrive* phase and this is shown in Fig. 2.11. As will be noted, the change in angular position of the variable displacement unit's wobble plate now causes it to pump high pressure oil to the fixed displacement unit. The cylinder block and input ring gear "housing" therefore rotate in the opposite direction to that of the underdrive phase, and so it increases the rotational speed of the planet gears and output ring gear "housing". Thus, speed is added to restore the required generator drive speed.

In multi-CSD generator systems the control of the drives is important in order that real electrical load (see p. 48) will be evenly distributed

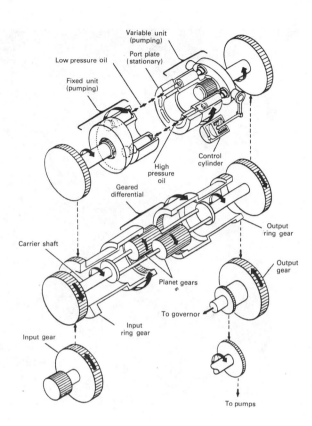

Fig 2.11
Overdrive phase

between the generators. Any unbalance in real load is automatically sensed by control units and load controllers in the generator systems and, since correction must be made at the generator drive, signals resulting from an unbalance are fed to an electromagnetic coil within the basic governor of each CSD (see page 39). The electromagnetic field interacts with additional permanent magnet flyweights driven by the governor, to produce a torque which in conjunction with centrifugal force provides a "fine" adjustment or trimming of the governor control valve, and of the output speed to the generators.

A typical CSD/generator installation is shown in Fig. 2.12.

The disconnection of a C.S.D. transmission system following a malfunction, may be accomplished mechanically by levers located in the flight crew compartment, electro-pneumatically, or as is more common, by an electro-mechanical system. In this system (see

Fig 2.12
CSD/generator installation

1. CSD	3. CSD oil service port	5. CSD oil cooler
2. Generator	4. CSD oil filter	6. Wet spline cavity service port

Fig. 2.13) the drive from the engine is transmitted to the C.S.D. via a dog-tooth clutch, and disconnect is initially activated by a solenoid controlled from the flight crew compartment.

When the solenoid is energized, a spring-loaded pawl moves into contact with threads on the input shaft and then serves as a screw causing the input shaft to move away from the input spline shaft (driven by the engine) thereby separating the driving dogs of the clutch. In some mechanisms a magnetically-operated indicator button is provided in the reset handle, which lies flush with the handle under normal operating conditions of the drive. When a disconnect has taken place, the indicator button is released from magnetic attraction and protrudes from the reset handle to provide a visual indication of the disconnect.

Resetting of disconnect mechanisms can only be accomplished on the ground following shutdown of the appropriate engine. In the system illustrated,

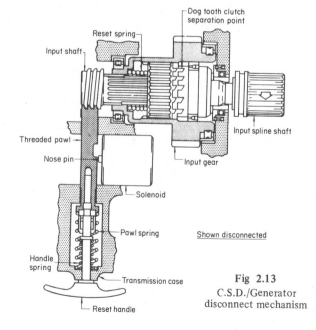

Fig 2.13
C.S.D./Generator
disconnect mechanism

resetting is accomplished by pulling out the reset handle to withdraw the threaded pawl from the input shaft, and allowing the reset spring on the shaft to re-engage the clutch. At the same time, and with the solenoid de-energized, the solenoid nose pin snaps into position in the slot of the pawl.

Generator Construction

A sectioned view of a typical constant frequency generator is illustrated in Fig. 2.14. It consists of three principal components: a.c. exciter which generates the power for the main generator field; rotating rectifier assembly mounted on, and rotating with, the rotor shaft to convert the exciter output to d.c.; and the main generator. All three components are contained within a cast aluminium casing made up of an end bell section and a stator

frame section; both sections are secured externally by screws. A mounting flange, which is an integral part of the stator frame, carries twelve slots reinforced by steel inserts, and key-hole shaped to facilitate attachment of the generator to the mounting studs of the constant-speed drive unit.

The exciter, which is located in the end bell section of the generator casing, comprises a stator and a three-phase star-wound rotor or exciter armature. The exciter armature is mounted on the same shaft as the main generator rotor and the output from its three-phase windings is fed to the rotating rectifier assembly.

The rotating rectifier assembly supplies excitation current to the main generator rotor field coils, and since together with the a.c. exciter they replace the conventional brushes and slip rings, they thereby eliminate the problems associated with them. The assembly is contained within a tubular insulator located in the hollow shaft on which the exciter and main generator

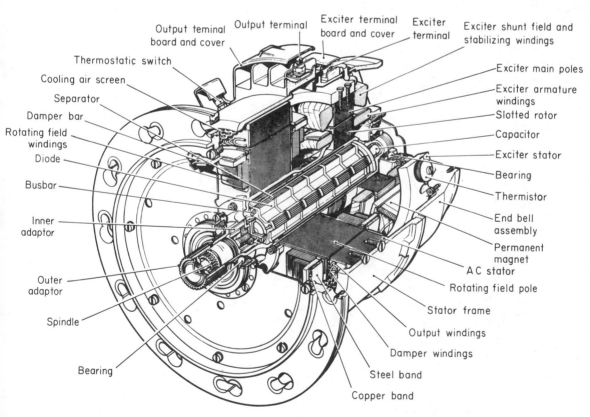

Fig 2.14
Constant frequency generator

rotors are mounted; located in this manner they are close to the axis of rotation and are not, therefore, subjected to excessive centrifugal forces. A suppression capacitor is also connected in the rectifier circuit and is mounted at one end of the rotor shaft. Its purpose is to suppress voltage "spikes" created within the diodes under certain operating conditions.

The main generator consists of a three-phase star-wound stator, and an eight-pole rotor and its associated field windings which are connected to the output of the rotating rectifier. The leads from the three stator phases are brought directly to the upper surface of an output terminal board, thus permitting the aircraft wiring to be clamped directly against the phase leads without current passing through the terminal studs. In addition to the field coils, damper (amortisseur) windings are fitted to the rotor and are located in longitudinal slots in the pole faces. Large copper bands, under steel bands at each end of the rotor stack, provide the electrical squirrel-cage circuit. The purpose of the damper windings is to provide an induction motor effect on the generator whenever sudden changes in load or driving torque tend to cause the rotor speed to vary above or below the normal or synchronous system frequency. In isolated generator operation, the windings serve to reduce excessively high transient voltages caused by line-to-line system faults, and to decrease voltage unbalance, during unbalanced load conditions. In parallel operation (see p. 47), the windings also reduce transient voltages and assist in pulling in, and holding, a generator in synchronism.

The drive end of the main rotor shaft consists of a splined outer adaptor which fits over a stub shaft secured to the main generator rotor. The stub shaft, in turn, fits over a drive spindle fixed by a centrally located screw to the hollow section of the shaft containing the rotating rectifier assembly. The complete shaft is supported at each end by pre-greased sealed bearings.

The generator is cooled by ram air which enters through the end bell section of the casing and passes through the windings and also through the rotor shaft to provide cooling of the rectifier assembly. The air is exhausted through a perforated screen around the periphery of the casing and at a point adjacent to the main generator stator. A thermally-operated overheat detector switch is screwed directly through the stator frame section into the stator of the main generator, and is connected to an overheat warning light on the relevant system control panel.

Further information on the circuit arrangement of the generator is given on page 45.

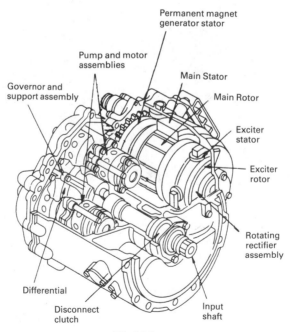

Fig 2.15
Integrated drive generator

Integrated Drive Generators

As will be noted from Fig. 2.15, an integrated drive
generator is one in which the CSD and generator
are mounted side by side to form a single compact
unit. This configuration reduces weight, requires
less space, and in comparison with the "end-to-end"
configuration it reduces vibration. The fundamental
construction and operation of both the generator
and drive units follow that described in the pre-
ceding paragraphs. The essential difference relates
to the method of cooling the generator. Instead of
air being utilized as the cooling medium, oil is
pumped through the generator; the oil itself is in
turn cooled by means of a heat exchanger system.

FIELD EXCITATION OF GENERATORS

The production of a desired output by any type of
generator requires a magnetic field to provide excitation of the windings for starting and for the subsequent operational running period. In other words,
a completely self-starting, self-exciting sequence is
required. In d.c. generators, this is achieved in a fairly

straightforward manner by residual magnetism in the
electromagnet system and by the build up of current
through the field windings. The field current, as it is
called, is controlled by a voltage regulator system.
The excitation of a.c. generators, on the other hand,
involves the use of somewhat more complex circuits,
the arrangements of which are essentially varied to
suit the particular type of generator and its controlling system. However, they all have one common
feature, i.e. the supply of direct current to the field
windings to maintain the desired a.c. output.

Frequency-Wild Generators

Figure 2.16 is a schematic illustration of the method
adopted for the generator illustrated in Fig. 2.7. In
this case, excitation of the rotor field is provided by
d.c. from the aircraft's main busbar and by rectified
a.c. The principal components and sections of the
control system associated with excitation are: the
control switch, voltage regulation section, field
excitation rectifier and current compounding section

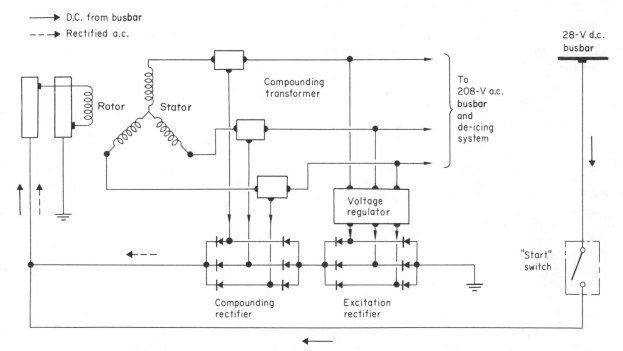

Fig. 2.16
Frequency-wild generator excitation

44

consisting of a three-phase current transformer and rectifier.

The primary windings of the compounding transformer are in series with the three phases of the generator and the secondary windings in series with the compounding rectifier.

When the control switch is in the "start" position, d.c. from the main busbar is supplied to the slip rings and windings of the generator rotor; thus, with the generator running, a rotating magnetic field is set up to induce an alternating output in the stator. The output is tapped to feed a magnetic amplifier type of voltage regulator which supplies a sensing current signal to the excitation rectifier (see p. 45). When this signal reaches a pre-determined off-load value, the rectified a.c. through the rotor winding is sufficient for the generator to become self-excited and independent of the main busbar supply which is then disconnected.

The maximum excitation current for wide-speed-range high-output generators of the type shown in Fig. 2.7 is quite high, and the variation in excitation current necessary to control the output under varying "load" conditions is such that the action of the voltage regulator must be supplemented by some other medium of variable excitation current. This is provided by the compounding transformer and rectifier, and by connecting them in the manner already described, direct current proportional to load current is supplied to the rotor field windings.

Constant-Frequency Generators

The exciter stator of the generator described on page 41 is made up of two shunt field windings, a stabilizing winding and also six permanent magnets; the latter provide a residual magnetic field for initial excitation. A thermistor is located in series with one of the parallel shunt field windings and serves as a temperature compensator. At low or normal ambient temperatures, the high resistance of the thermistor blocks current flow in its winding circuit so that it causes the overall shunt field resistance to be about that of the remaining winding circuit. At the higher temperature resulting from normal operation, the resistance of each single circuit increases to approximately double. At the same time, however, the thermistor resistance drops to a negligible value permitting approximately equal current to flow in each winding circuit.

The stabilizing winding is wound directly over the shunt field windings, and with the permanent magnet poles as a common magnetic core, a transformer type of coupling between the two windings is thereby provided. The rectifier assembly consists of six silicon diodes separated by insulating spacers and connected as a three-phase full-wave bridge.

The excitation circuit arrangement for the generator is shown schematically in Fig. 2.17. When the generator starts running, the flux from the permanent magnets of the a.c. exciter provides the initial flow of current in its rotor windings. As a result of the initial current flow, armature reaction is set up and owing to the position of the permanent magnetic poles, the reaction polarizes the main poles of the exciter stator in the proper direction to assist the voltage regulator in taking over excitation control.

The three-phase voltage produced in the windings is supplied to the rectifier assembly, the d.c. output of which is, in turn, fed to the field coils of the main generator rotor as the required excitation current. A rotating magnetic field is thus produced which induces a three-phase voltage output in the main stator windings. The output is tapped and is fed back to the shunt field windings of the exciter, through the voltage regulator system, in order to produce a field supplementary to that of the permanent magnets. In this manner the exciter output is increased and the main generator is enabled to build up its output at a faster rate. When the main output reaches the rated value, the supplementary electromagnetic field controls the excitation and the effect of the permanent magnets is almost eliminated by the opposing armature reaction. During the initial stages of generator operation, the current flow to the exciter only passes through one of the two shunt field windings, due to the inverse temperature/resistance characteristics of the thermistor. As the temperature of the winding increases, the thermistor resistance decreases to allow approximately equal current to flow in both windings, thus maintaining a constant effect of the shunt windings.

In the event that excitation current should suddenly increase or decrease as a result of voltage fluctuations due, for example, to switching of loads, a current will be induced in the stabilizing winding since it acts as a transformer secondary winding. This current is fed into the voltage regulator as a feedback signal to so adjust the excitation current that voltage fluctuations resulting from any cause are opposed and held to a minimum.

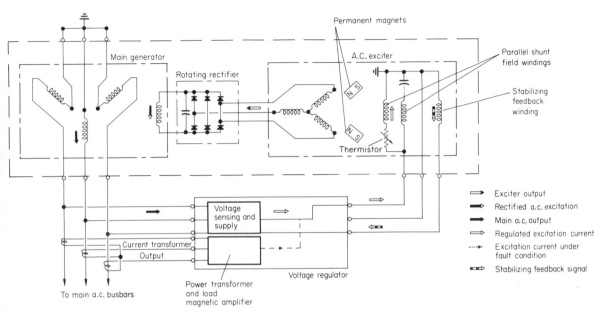

Permanent magnets

Main generator

A.C. exciter

Rotating rectifier

Parallel shunt
field windings

Stabilizing
feedback
winding

Thermistor

Voltage
sensing and
supply

Current transformer

Output

Power transformer
and load
magnetic amplifier

Voltage regulator

To main a.c. busbars

⇒ Exciter output

➡ Rectified a.c. excitation

➡ Main a.c. output

⇨ Regulated excitation current

--→ Excitation current under
fault condition

▭▶ Stabilizing feedback signal

Fig 2.17
Circuit diagram of constant frequency generator

VOLTAGE REGULATION OF GENERATORS

The control of the output voltages of a.c. generators is also an essential requirement, and from the foregoing description of excitation methods, it will be recognized that the voltage regulation principles adopted for d.c. generators can also be applied, i.e. automatic adjustment of excitation current to meet changing conditions of load and/or speed. Voltage regulators normally form part of generator system control and protection units.

Frequency-Wild Generators

Figure 2.18 is a block functional diagram of the method used for the voltage regulation of the generator illustrated in Fig. 2.7. Regulation is accomplished by a network of magnetic amplifiers or transducers, transformers and bridge rectifiers interconnected as shown. In addition to the control of load current delivered by the generator, a further factor which will affect control of field excitation is the error between the line voltage desired and the actual voltage obtained. As already explained on page 44, the compounding transformer and rectifier provides excitation current proportional to load current, therefore the sensing of error voltages and necessary re-adjustment of excitation current must be provided by the voltage regulation network.

It will be noted from the diagram that the three-phase output of the generator is tapped at two points; at one by a three-phase transformer and at the other by a three-phase magnetic amplifier. The secondary winding of one phase of the transformer is connected to the a.c. windings of a single-phase "error sensing" magnetic amplifier and the three primary windings are connected to a bridge "signal" rectifier. The d.c. output from the rectifier is then fed through a voltage-sensing circuit made up of two resistance arms, one (arm "A") containing a device known as a barretter the characteristics of which maintain a substantially constant current through the arm, the other (arm "B") of such resistance that the current flowing through it varies linearly with the line voltage. The two current signals, which are normally equal at the desired line voltage, are fed in opposite directions over the a.c. output windings in the error magnetic amplifier. When there is a change in the voltage level, the resulting variation in current flowing through arm "B" unbalances the sensing circuit and, as this circuit has the same function as a d.c. control winding, it changes the reactance of the error magnetic amplifier a.c. output windings and an amplified error signal current is produced. After rectification, the signal is then fed as d.c. control current to the three-phase magnetic amplifier, thus causing its reactance and a.c. output to change

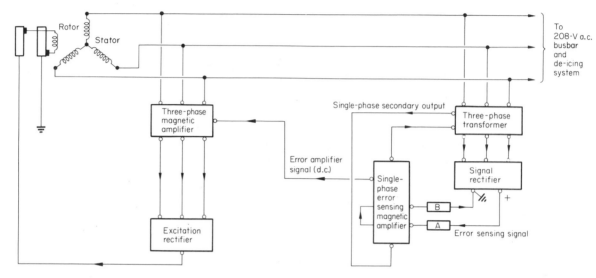

Rotor
Stator
To 208-V a.c. busbar and de-icing system
Three-phase magnetic amplifier
Single-phase secondary output
Three-phase transformer
Error amplifier signal (d.c.)
Signal rectifier
Single-phase error sensing magnetic amplifier
B
A
+
Error sensing signal
Excitation rectifier

Fig 2.18
Voltage regulation

also. This results in an increase or decrease, as appropriate, of the excitation current flow to the generator rotor field winding, continuing until the line voltage produces balanced signal conditions once more in the error sensing circuit.

Constant-Frequency Systems

The regulation of the output of a constant-frequency system is also based on the principle of controlling field excitation, and some of the techniques thus far described are in many instances applied. In installations requiring a multi-arrangement of constant-frequency generators, additional circuitry is required to control output under load-sharing or parallel operating conditions and as this control also involves field excitation, the overall regulation circuit arrangement is of an integrated, and sometimes complex, form. At this stage, however, we are only concerned with the fundamental method of regulation and for this purpose we may consider the relevant sections or stages of the circuit shown schematically in Fig. 2.19.

The circuit is comprised of three main sections: a voltage error detector, pre-amplifier and a power amplifier. The function of the voltage error detector is to monitor the generator output voltage, compare it with a fixed reference voltage and to transmit any error to the pre-amplifier. It is made up of a three-phase bridge rectifier connected to the generator output, and a bridge circuit of which two arms contain

gas-filled regulator tubes and two contain resistances. The inherent characteristics of the tubes are such that they maintain an essentially constant voltage drop across their connections for a wide range of current through them and for this reason they establish the reference voltage against which output voltage is continuously compared. The output side of the bridge is connected to an "error" control winding of the pre-amplifier and then from this amplifier to a "signal" control winding of a second stage or power amplifier. Both stages are three-phase magnetic amplifiers. The final amplified signal is then supplied to the shunt windings of the generator a.c. exciter stator (see also Fig. 2.17).

The output of the bridge rectifier in the error detector is a d.c. voltage slightly lower than the average of the three a.c. line voltages; it may be adjusted by means of a variable resistor (RV_1) to bring the regulator system to a balanced condition for any nominal value of line voltage. A balanced condition of the bridge circuit concerned is obtained when the voltage applied across the bridge (points "A" and "B") is exactly twice that of the voltage drop across the two tubes. Since under this condition, the voltage drop across resistors R_1 and R_2 will equal the drop across each tube, then no current will flow in the output circuit to the error control winding of the pre-amplifier.

If the a.c. line voltage should go above or below the fixed value, the voltage drops across R_1 and R_2 will

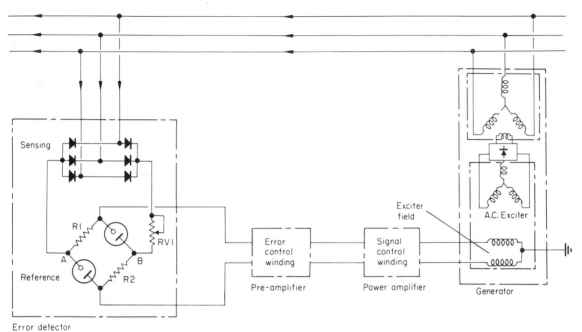

Fig 2.19

Constant-frequency system voltage regulation

differ causing an unbalance of the bridge circuit and a flow of current to the "error" control winding of the pre-amplifier. The direction and magnitude of current flow will depend on whether the variation, or error in line voltage, is above (positive error signal) or below (negative error signal) the balanced nominal value, and on the magnitude of the variations.

When current flows through the "error" control winding the magnetic flux set up alters the total flux in the cores of the amplifier, thereby establishing a proportional change in the amplifier output which is applied to the signal winding of the power amplifier. If the error signal is negative it will cause an increase in core flux, thereby increasing the power amplifier output current to the generator exciter field winding. For a positive error signal the core flux and excitation current will be reduced. Thus, the generator output is controlled to the preset value which on being attained restores the error detector bridge circuit to the balanced condition.

Regulators normally incorporate torque-limiting circuitry which limits the torque at mechanical linkages to a safe value by limiting the exciter field current.

LOAD-SHARING OR PARALLELING

Frequency-Wild Systems

In systems of this type, the a.c. output is supplied to independent consumer equipment and since the frequency is allowed to go uncontrolled, then paralleling or sharing of the a.c. load is not possible. In most applications this is by design; for example, in electrical de-icing equipment utilizing resistance type heaters, a variable frequency has no effect on system operation; therefore reliance is placed more on generator dependability and on the simplicity of the generating system. In rectified a.c. systems frequency is also uncontrolled, but as most of the output is utilized for supplying d.c. consumer equipment, load sharing is more easily accomplished by paralleling the rectified output through equalizing circuits in a similar manner to that adopted for d.c. generating systems (see p. 16).

Constant-Frequency Systems

These systems are designed for operation under load-sharing or paralleling conditions and in this connection regulation of the two parameters, *real load* and *reactive load*, is required. Real load is the actual working load output in kilowatts (kW) available for supplying the various electrical services, and the reactive load is the so-called "wattless load" which is in fact the vector sum of the inductive and capacitive currents and voltage in the system expressed in kilovolt-amperes reactive (kVAR). (See Fig. 2.6 once again.)

Since the real load is directly related to the input power from the prime mover, i.e. the aircraft engine, real load-sharing control must be on the engine. There are, however, certain practical difficulties involved, but as it is possible to reference back any real load unbalance to the constant-speed drive unit between engine and generator, real load-sharing control is effected at this unit by adjusting torque at the output drive shaft.

Reactive load unbalances are corrected by controlling the exciter field current delivered by the voltage regulators to their respective generators, in accordance with signals from a reactive load-sharing circuit.

Real Load-Sharing   The sharing of real load between paralleled generators is determined by the real relative rotational speeds of the generators which in turn influence the voltage phase relationships.

As we learned earlier (see p. 38) the speed of a generator is determined by the initial setting of the governor on its associated constant speed drive. It is not possible, however, to obtain exactly identical governor settings on all constant speed drives employed in any one installation, and so automatic control of the governors becomes necessary.

A.C. generators are synchronous machines. Therefore when two or more operate in parallel they lock together with respect to frequency and the system frequency established is that of the generator whose output is at the highest level. Since this is controlled by speed-governing settings then it means that the generator associated with a higher setting will carry more than its share of the load and will supply energy which tends to motor the other machines in parallel with it. Thus, sharing of the total real load is unbalanced, and equal amounts of energy in the form of torque on the generator rotors must be supplied.

Fundamentally, a control system is comprised of two principal sections: one in which the unbalance is determined by means of current transformers, and the other (load controlling section) in which torques are established and applied. A circuit diagram of the system as applied to a four-generator installation is shown schematically in Fig. 2.20.

The current transformers sense the real load distribution at phase "C" of the supply from each generator, and are connected in series and together they form a load sharing loop. Each load controller is made up of a two-stage magnetic amplifier controlled by an error sensing element in parallel with each current transformer. The output side of each load controller is, in turn, connected to a solenoid in the speed governor of each constant speed unit.

When current flows through phase "C" of each generator, a voltage proportional to the current is induced in each of the current transformers and as they are connected in series, then current will flow in the load sharing loop. This current is equal to the average of the current produced by all four transformers.

Let us assume that at one period of system operation, balanced load sharing conditions are obtained under which the current output from each transformer is equal to five amps, then the average flowing in the load sharing loop will be five amps, and no current circulates through the error sensing elements. If now a generator, say No. 1, runs at a higher speed governor setting than the other three generators, it will carry more load and will increase the output of its associated current transformer.

The share of the load being carried by the other generators falls proportionately, thereby reducing the output of their current transformers, and the average current flowing in the load sharing loop remains the same, i.e. five amps. If, for example, it is assumed that the output of No. 1 generator current transformer is increased to eight amps a difference of three amps will flow through the error sensing element of its relevant load controller. The three amps difference divides equally between the other generators and so the output of each corresponding current transformer is reduced by one amp, a difference which flows through the error sensing elements of the load controllers. The error signals are then applied as d.c. control signals to the two-stage magnetic amplifiers and are fed to electromagnetic coils which are mounted adjacent to permanent magnet flyweights and form part of the governor in each constant speed drive unit (see page 39). The current and magnetic field simulate the effects of centrifugal forces on the flyweights and are of such direction and magnitude as to cause the flyweights to be attracted or repelled.

Thus, in the unbalanced condition we have assumed,

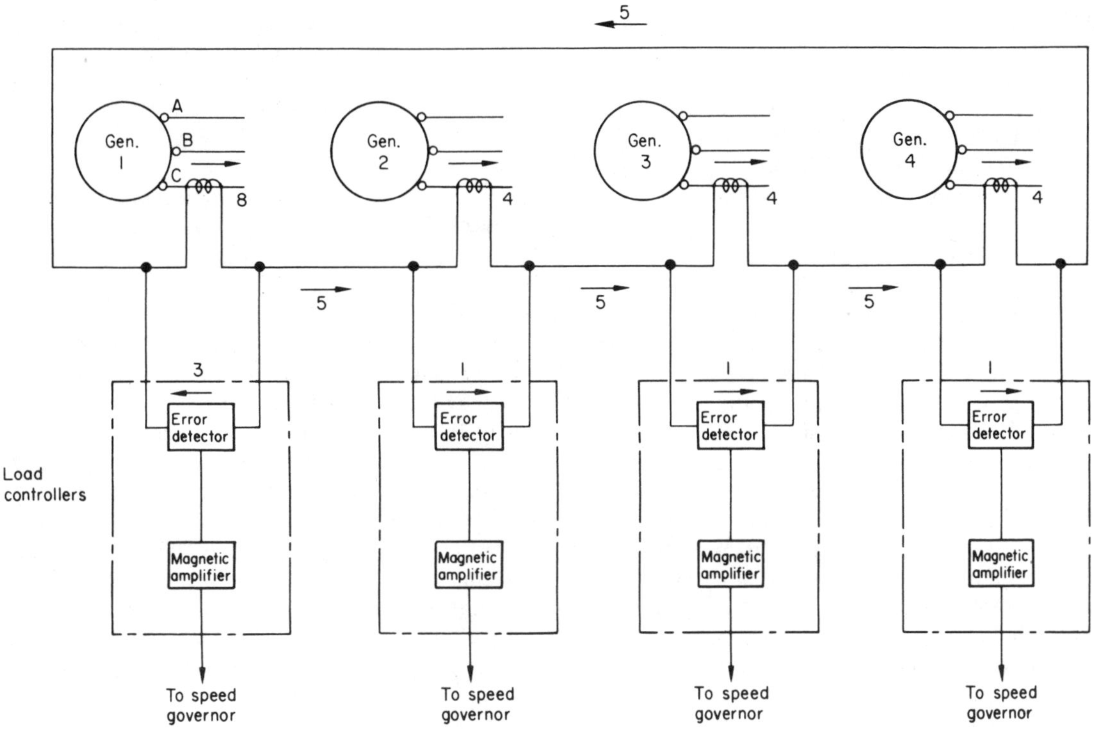

Fig 2.20
Real load-sharing

i.e. No. 1 generator running at a higher governor setting, the current and field resulting from the error
signal applied to the corresponding load controller
flows in the opposite sense and repels the flyweights,
thereby simulating a decrease of centrifugal force.
The movement of the flyweights causes oil to flow to
underdrive and the output speed of the constant speed
unit drive decreases, thereby correcting the governor
setting to decrease the load being taken by No. 1
generator. The direction of the current and field in
the load controller sensing elements of the remaining
generators is such that the governor flyweights in
their constant speed drive units are attracted, allowing
oil to flow to overdrive, thereby increasing the load
being taken by each generator.

Reactive Load-Sharing The sharing of reactive load
between paralleled generators depends on the relative
magnitudes of their output voltages which vary, and
as with all generator systems are dependent on the
settings of relevant voltage regulators and field
excitation current (see also p. 43). If, for example,

the voltage regulator of one generator is set slightly
above the mean value of the whole parallel system,
the regulator will sense an under-voltage condition
and it will accordingly increase its excitation current
in an attempt to raise the whole system voltage to
its setting. However, this results in a reactive component of current flowing from the "over-excited"
generator which flows in opposition to the reactive
loads of the other generators. Thus, its load is
increased while the loads of the other generators
are reduced and unbalance in reactive load sharing
exists. It is therefore necessary to provide a circuit to
correct this condition.

In principle, the method of operation of the reactive
load-sharing circuit is similar to that adopted in the
real load-sharing circuit described earlier. A difference
in the nature of the circuitry should however be noted
at this point. Whereas in the real load-sharing circuit
the current transformers are connected directly to
the error detecting elements in load controlling units,
in a reactive load-sharing circuit (see Fig. 2.21) they
are connected to the primary windings of devices

50

called mutual reactors. These are, in fact, transformers which have (i) a power source connected to their secondary windings in addition to their primaries; in this instance, phase "C" of the generator output, and (ii) an air gap in the iron core to produce a phase displacement of approximately 90 degrees between the primary current and secondary voltage. They serve the purpose of delivering signals to the voltage regulator which is proportional to the generator's reactive load only.

When a reactive load unbalance occurs, the current transformers detect this in a similar manner to those associated with the real load-sharing circuit and they cause differential currents to flow in the primary windings of their associated mutual reactors. Voltages proportional to the magnitude of the differential currents are induced in the secondary windings and will either lead or lag generator current by 90 degrees. When the voltage induced in a particular reactor secondary winding leads the associated generator current it indicates that a reactive load exists on the generator; in other words, that it is taking more than its share of the total load. In this condition, the

voltage will add to the voltage sensed by the secondary winding at phase "C". If, on the other hand, the voltage lags the generator current then the generator is absorbing a reactive load, i.e. it is taking less share of the total load and the voltage will subtract from that sensed at phase "C".

The secondary winding of each mutual reactor is connected in series with an error detector in each voltage regulator, the detector functioning in the same manner as those used for voltage regulation and real load-sharing (see pp. 48 and 49).

Let us assume that No. 1 generator takes the greater share of the load, i.e. it has become over-excited. The voltage induced in the secondary winding of the corresponding mutual reactor will be additive and so the error detector will sense this as an overvoltage. The resulting d.c. error signal is applied to the pre-amplifier and then to the power amplifier the output of which is adjusted to reduce the amount of exciter current being delivered to the No. 1 generator. In the case of the other three generators they will have been carrying less than their share of the reactive load and, therefore, the

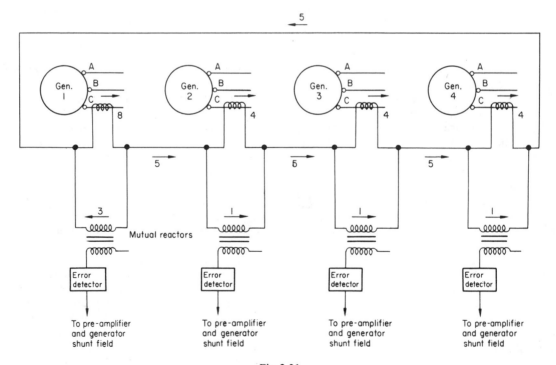

Fig 2.21
Reactive load-sharing

voltages induced in their mutual reactors will have lagged behind the currents from the generators, resulting in opposition to the voltages sensed by the secondary windings. Thus, the output of each power amplifier will be adjusted to increase the amount of exciter current being delivered to their associated generators until equal reactive load-sharing is restored between generators within the prescribed limits.

Synchronizing Lights

In some power-generating systems a method of indicating synchronization between generator outputs forms part of the paralleling system, and consists of lights and frequency adjustment controls.
A schematic diagram of the method based on that adopted for the triple generator system of the Boeing 727 is given in Fig. 2.22.

The lights are connected into phases "A" and "C" of each generator between the generator breakers and a synchronizing busbar, via a selector switch. The switch is also used for connecting a voltmeter and a frequency meter to each generator output phase "B". The frequency adjustment controls are connected into the circuit of the load controllers (see also page 49).

The generators are connected to their respective load busbars and the synchronizing busbar, via generator breakers and bus-tie breakers respectively; each breaker being closed or tripped by manual operation of switches on a panel at the Flight Engineer's station. The breakers also trip automatically in the event of faults detected by the generator control and protection system. The field relays are similarly operated. Indicator lights are located adjacent to all switches to indicate either of the closed or tripped conditions.

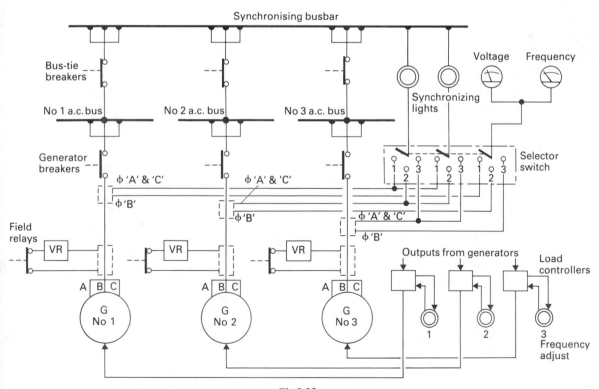

Fig 2.22
Synchronizing

Prior to engine starting, the bus-tie breakers and field relays are closed (indicator lights out) and the generator breakers are tripped (indicator lights on). As the first engine is started, the meter selector switch is positioned at GEN 1 to connect phases "A" and "C" of this generator to the synchronizing busbar via the synchronizing lights. Phase "B" is connected to both the voltmeter and frequency meter the readings of which are then checked. Since at this moment, only the number 1 generator is in operation, then with respect to the other two, it will of course, produce maximum voltage and phase difference and both synchronizing lights will flash at a high frequency as a result of the current flow through them. The frequency control knob for the generator is then adjusted until its load controller has trimmed the CSD/generator speed to produce a "master" frequency of about 403 Hz, and simultaneous flashing of both synchronizing lights.

When the second engine is started, the meter selector switch is positioned at GEN 2 to connect the synchronizing lights and meters to the appropriate phases of number 2 generator, and its frequency is also adjusted in the manner just described. The number 1 generator is then connected to its load busbar by closing its generator breaker. This action also connects the generator to the synchronizing busbar, and since the synchronizing lights are now sensing the output of the second on-coming generator, their flashing frequency will be very much less as a result of less voltage and phase difference between the two generator outputs. The frequency of the second generator is then adjusted to obtain the greatest time interval between flashes of the synchronizing lights, and while the lights are out (indicating both sources of power are in phase) the number 2 generator is connected to its load busbar by closing its breaker.

As the third engine is started, the meter selector switch is positioned at GEN 3, and by following the same procedure just outlined, number 3 generator is connected to its load busbar. With all three generators thus connected their subsequent operation is taken care of automatically by the load-sharing sensing circuits of the associated control and protection unit.

It is important to note that a generator must never be connected to its load busbar when the synchronizing lights are on. Such action would impose heavy loads on the generator or CSD and possibly cause damage to them. If, at any time the synchronizing lights flash alternately, a phase reversal is indicated and the appropriate generator should not be used.

AIR-DRIVEN GENERATORS

The application of generators dependent upon an airstream as the prime mover is by no means a new one and, having been adopted in many early types of aircraft for the generation of electrical power, the idea of repeating the practice for to-day's advanced electrical systems would, therefore, seem to be retrogressive. However, an air-drive can serve as a very useful stand-by in the event of failure of a complete main a.c. generating system and it is in this emergency role that it is applied to some types of aircraft.

The drive consists of a two-bladed fan or air turbine as it is sometimes called, and a step-up ratio gear train which connects the fan to a single a.c. generator. The generator is of a similar type to the main generator (see also p. 41) but has a lower output rating since it is only required to supply the consumer equipment essential under emergency conditions. The complete unit is stowed on a special mounting in the aircraft fuselage, and when required is deployed by a mechanically linked release handle in the flight compartment. When deployed at airspeeds of between 120 to 430 knots, the fan and generator are driven up to their appropriate speeds by the airstream, and electrical power is delivered via a regulator at the rated values. A typical nominal fan speed is 4,800 rev/min and is self-governed by varying the blade pitch angles. The gearbox develops a generator shaft speed of 12,000 rev/min. After deployment of the complete unit, it can only be restowed when the aircraft is on the ground.

CHAPTER THREE

Power Conversion Equipment

In aircraft electrical installations a number of different types of consumer equipment are used which require power supplies different from those standard supplies provided by the main generator. For example, in an aircraft having a 28 volts d.c. primary power supply, certain instruments and electronic equipment are employed which require 26 volts and 115 volts a.c. supplies for their operation, and as we have already seen, d.c. cannot be entirely eliminated even in aircraft which are primarily a.c. in concept. Furthermore, we may also note that even within the items of consumer equipment themselves, certain sections of their circuits require different types of power supply and/or different levels of the same kind of supply. It therefore becomes necessary to employ not only equipment which will convert electrical power from one form to another, but also equipment which will convert one form of supply to a higher or lower value.

The equipment required for the conversion of main power supplies can be broadly divided into two main types, static and rotating, and the fundamentals of construction and operation of typical devices and machines are described under these headings.

Static Converting Equipment

The principal items which may be grouped under this heading are rectifiers and transformers, some applications of which have already been discussed in Chapter 2, and static d.c./a.c. converters.

The latter items are transistorized equivalents of rotary inverters and a description of their construction and operating fundamentals will be given at the end of this chapter.

RECTIFIERS

The process of converting an a.c. supply into a d.c. supply is known as rectification and any static apparatus used for this purpose is known as a rectifier.

The rectifying action is based on the principle that when a voltage is applied to certain combinations of metallic and non-metallic elements in contact with each other, an exchange of electrons and positive current carriers (known as "holes") takes place at the contact surfaces. As a result of this exchange, a barrier layer is formed which exhibits different resistance and conductivity characteristics and allows current to flow through the element combination more easily in one direction than in the opposite direction. Thus, when the applied voltage is an alternating quantity the barrier layer converts the current into a undirectional flow and provides a rectified output.

One of the elements used in combination is referred to as a "semi-conductor" which by definition denotes that it possesses a resistivity which lies between that of a good conductor and a good insulator. Semi-conductors are also further defined by the number of carriers, i.e. electrons and positive "holes", provided by the "crystal lattice" form of the element's atomic structure. Thus, an element having a majority of electron carriers is termed "n-type" while a semi-conductor having a majority of "holes" is termed "p-type".

If a p-type semi-conductor is in contact with a metal plate as shown in Fig. 3.1, electrons migrate from the metal to fill the positive holes in the semi-conductor, and this process continues until the transference of charge has established a p.d. sufficient to stop it. By this means a very thin layer of the semi-conductor is cleared of positive holes and thus becomes an effective insulator, or barrier layer. When a voltage is applied such that the semi-conductor is positive with respect to the metal, positive holes migrate from the body of the semi-conductor into the barrier layer, thereby reducing its "forward" resistance and restoring con-

54

ductivity. If, on the other hand, the semi-conductor is made negative to the metal, further electrons are drawn from the metal to fill more positive holes and the "reverse" resistance of the barrier layer is thus increased. The greater the difference in the resistance to current flow in the two directions the better is the rectifying effect.

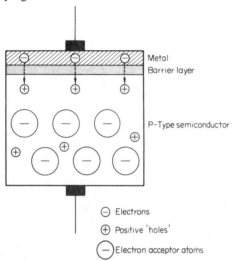

Fig. 3.1
Semi-conductor/metal junction

A similar rectifying effect is obtained when an n-type semi-conductor is in contact with metal and a difference of potential is established between them, but in this case the direction of "easy" current flow is reversed. In practice, a small current does flow through a rectifier in the reverse direction because p-type material contains a small proportion of free electrons and n-type a small number of positive holes.

In the rectification of main a.c. power supplies, rectifiers are now invariably of the type employing the p-type non-metallic semi-conductors, selenium and silicon. Rectifiers employing germanium (a metallic element) are also available but as their operating temperature is limited and protection against short duration overloads is difficult, they are not adopted in main power systems.

SELENIUM RECTIFIERS
The selenium rectifier is formed on an aluminium sheet which serves both as a base for the rectifying junction and as a surface for the dissipation of heat. A cross-section of an element is shown diagrammatically in Fig. 3.2 and from this it will be noted that

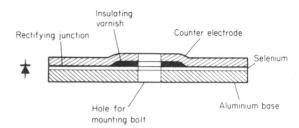

Fig 3.2
Cross-section of a selenium rectifier element

the rectifying junction covers one side of the base with the exception of a narrow strip at the edges and a small area around the fixing hole which is sprayed with a layer of insulating varnish. A thin layer of a low-melting point alloy, referred to as the counter electrode, is sprayed over the selenium coating and insulating varnish. Contact with the two elements of the rectifying junction, or barrier layer, is made through the base on one side and the counter electrode on the other.

Mechanical pressure on the rectifying junction tends to lower the resistance in the reverse direction and this is prevented in the region of the mounting studs by the layer of varnish.

In practice a number of rectifying elements may be connected in series or parallel to form what is generally referred to as a rectifier stack. Two typical stacks are shown in Fig. 3.3.

Fig 3.3
Typical rectifier stacks

When connected in series the elements increase the voltage handling ability of a rectifier and when connected in parallel the ampere capacity is increased.

SILICON RECTIFIERS
Silicon rectifiers, or silicon junction diodes as they are commonly known, do not depend on such a large barrier layer as selenium rectifiers, and as a result they differ radically in both appearance and size. This will

be apparent from Fig. 3.4 which illustrates a junction diode of a type similar to that used in the constant-frequency generator described in Chapter 2.

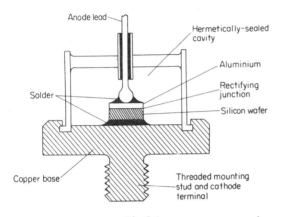

Fig 3.4
Silicon junction diode

The silicon is in the form of an extremely small slice cut from a single crystal and on one face it has a fused aluminium alloy contact to which is soldered an anode and lead. The other face is soldered to a base, usually copper, which forms the cathode and at the same time serves as a heat sink and dissipator. The barrier layer is formed at the aluminium-silicon junction.

To protect the junction from water vapour and other deleterious materials, which can seriously impair its performance, it is mounted in a hermetically-sealed case.

OPERATING LIMITATIONS OF RECTIFIERS

The limiting factors in the operation of a rectifier are: (i) the maximum temperature permissible and (ii) the minimum voltage, i.e. the reverse voltage, required to break down the barrier layer. In selenium rectifiers the maximum temperature is of the order of 70°C. For germanium the temperature is about 50°C, while for silicon up to 150°C may be reached without destroying the rectifier. It should be noted that these figures represent the actual temperature at the rectifying junction and therefore the rectifier, as a complete unit, must be at a much lower temperature. Proper cooling under all conditions is, therefore, an essential requirement and is normally taken care of by blower motors or other forced air methods such as the one adopted in the constant-frequency generator referred to earlier.

Voltage ratings are determined by the ability of a rectifier to withstand reverse voltage without passing excessive reverse current, and the characteristics are such that reverse current does not increase proportionately to the applied voltage. This is because once all the current carriers have been brought into action there is nothing to carry any further current. However, at a sufficiently high voltage the resistance in the reverse direction breaks down completely and reverse current increases very sharply. The voltage at which breakdown occurs is called the Zener voltage, and as it depends on the impurity content of the material used, a constant value can be chosen by design and during manufacture of a rectifier. For power rectification, rectifiers must have a high Zener voltage value and each type must operate at a reverse voltage below its designed breakdown value. Some rectifiers, however, are designed to break down at a selected value within a low voltage range (between 2 and 40 volts is typical) and to operate safely and continuously at that value. These rectifiers are called Zener Diodes and since the Zener voltage is a constant and can therefore serve as a reference voltage, they are utilized mostly in certain low voltage circuits and systems for voltage level sensing and regulation (see also p. 15).

SILICON CONTROLLED RECTIFIER (S.C.R.)

An S.C.R., or thyristor as it is sometimes called, is a development of the silicon diode and it has some of the characteristics of a thyratron tube. It is a three-terminal device, two terminals corresponding to those of an ordinary silicon diode and the third, called the "gate" and corresponding to the thyratron grid. The construction and operating characteristics of the device are shown in Fig. 3.5. The silicon wafer which is of the "n-type" has three more layers formed within it in the sequence indicated.

When reverse voltage is applied an S.C.R. behaves in the same manner as a normal silicon diode, but when forward voltage is applied current flow is practically zero until a forward critical "breakover" voltage is reached. The voltage at which breakover takes place can be varied by applying small current signals between the gate and the cathode, a method known as "firing". Once conduction has been initiated it can be stopped only by reducing the voltage to a very low value. The mean value of rectified voltage can be controlled by adjusting the phasing of the gate signal with respect to the applied voltage. Thus, an S.C.R. not only performs the function of power rectification, but also the function of an on-off switch, and a vari-

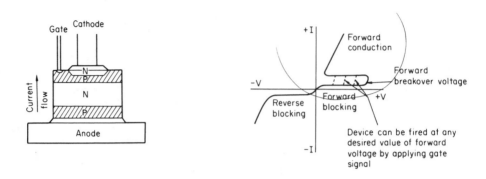

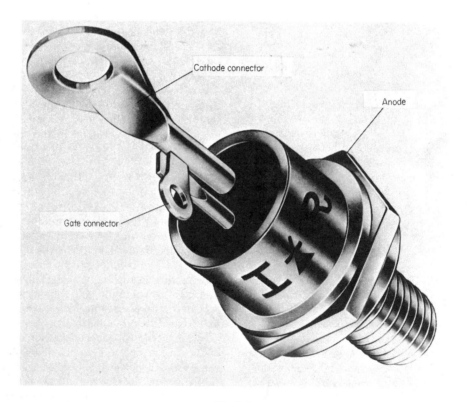

Fig. 3.5
Silicon controlled rectifier

able power output device. A typical application of S.C.R. switching is in the battery charger unit already referred to on p. 29. Fig. 3.6 shows how an S.C.R.

produces a variable d.c. voltage which, for example, would be required in a variable speed motor circuit, as gate signal currents or "firing point" is varied.

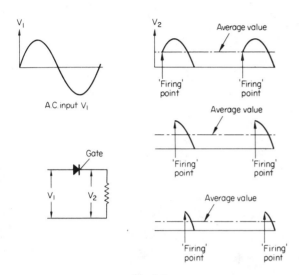

Fig. 3.6
Variable d.c. output from a silicon controlled rectifier

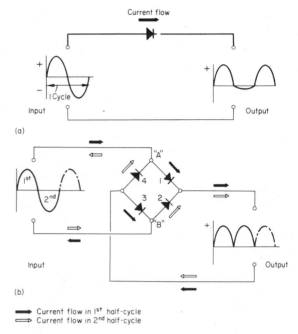

➡ Current flow in 1ˢᵗ half-cycle
⇨ Current flow in 2ⁿᵈ half-cycle

Fig. 3.7
Single-phase rectification
(a) Half-wave
(b) Full-wave

RECTIFIER CIRCUIT CONNECTIONS

Rectifiers are used in single-phase and three-phase supply systems and, depending on the conversion requirements of a circuit or system, they may be arranged to give either half-wave or full-wave rectification. In the former arrangement the d.c. output is available only during alternate half-cycles of an a.c. input, while in the latter a d.c. output is available throughout a cycle.

The single-phase half-wave circuit shown in Fig. 3.7(a) is the simplest possible circuit for a rectifier and summarizes, in a practical manner, the operating principles already described. The output from the single rectifier is a series of positive pulses the number of which is equal to the frequency of the input voltage. For a single-phase a.c. input throughout a full cycle, a bridge connection of rectifiers is used (Fig. 3.7(b)).

For half-wave rectification of a three-phase a.c. input the circuit is made up of three rectifiers in the manner shown in Fig. 3.8. This arrangement is comparable to three single-phase rectification circuits, but since the positive half-cycles of the input are occurring at time intervals of one third of a cycle (120 degrees) the number of d.c. pulses or the ripple frequency is increased to three times that of the supply and a smoother output waveform is obtained.

Figure 3.9 shows the circuit arrangement for the full-wave rectification of a three-phase a.c. input; it

is of the bridge type and is most commonly used for power rectification in aircraft. Examples of three-phase bridge rectifier applications have already been shown in Chapter 2 but we may now study the circuit operations in a little more detail.

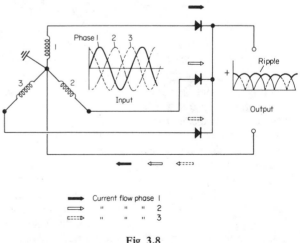

➡ Current flow phase 1
⇨ " " " 2
⇨ " " " 3

Fig. 3.8
Three-phase half-wave rectification

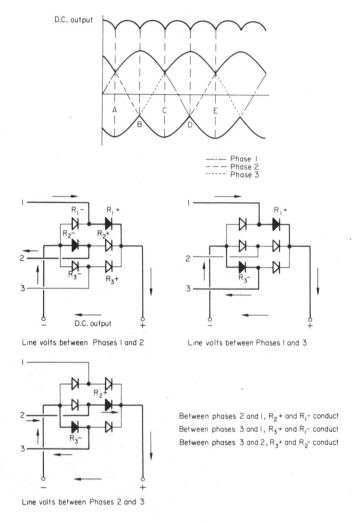

Fig 3.9
Operation of a full-wave bridge rectifier

In this type of circuit only two rectifiers are conducting at any instant; one on the positive side and the other on the negative side. Also the voltage applied to the bridge network is that between two of the phases, i.e. the line voltage. Let us consider the points "A" and "B" on the three phase voltage curves. These points represent the line voltage between phases 1 and 2 of the supply and from the circuit diagram we note that rectifiers $R_1 +$ and $R_2 -$ only will conduct. From "B" to "C" the line voltage corresponds to that between phases 1 and 3 and $R_1 +$ now conducts in conjunction with $R_3 -$. Between the points "C" and "D" the line voltage corresponds to that between phases 2 and 3 so that rectifier $R_2 +$ now takes over and conducts in conjunction with $R_3 -$. This process continues through the remaining three conducting paths, the sequence of the relevant phases and the rectifiers which conduct being as tabulated in Fig. 3.9.

The output voltage, which is determined by the distance between the positive and negative crests, consists of the peaks of the various line voltages for phase angles of 30 degrees on either side of their maxima. Since the negative half-cycles are included, then the ripple frequency of a bridge rectifier output is six times that of the a.c. input and an even smoother waveform is obtained.

TRANSFORMERS

A transformer is a device for converting a.c. at one frequency and voltage to a.c. at the same frequency but at another voltage. It consists of three main parts: (i) an iron core which provides a circuit of low reluctance for an alternating magnetic field created by, (ii) a primary winding which is connected to the main power source and (iii) a secondary winding which receives electrical energy by mutual induction from the primary winding and delivers it to the secondary circuit. There are two classes of transformers, voltage or power transformers and current transformers.

Principle. The three main parts are shown schematically in Fig. 3.10. When an alternating voltage is applied to the primary winding an alternating current will flow and by self-induction will establish

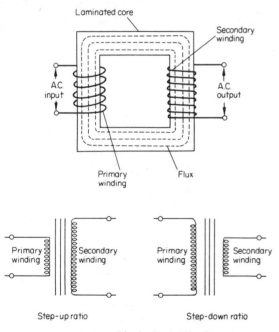

Fig 3.10
Transformer principle

a voltage in the primary winding which is opposite and almost equal to the applied voltage. The difference between these two voltages will allow just enough current (excitation current) to flow in the primary winding to set up an alternating magnetic flux in the core. The flux cuts across the secondary winding and by mutual induction (in practice both windings are

wound one on the other) a voltage is established in the secondary winding.

When a load is connected to the secondary winding terminals, the secondary voltage causes current to flow through the winding and a magnetic flux is produced which tends to neutralize the magnetic flux produced by the primary current. This, in turn, reduces the self-induced, or opposition, voltage in the primary winding, and allows more current to flow in it to restore the core flux to a value which is only very slightly less than the no-load value.

The primary current increases as the secondary load current increases, and decreases as the secondary load current decreases. When the load is disconnected, the primary winding current is again reduced to the small excitation current sufficient only to magnetize the core.

To accomplish the function of changing voltage from one value to another, one winding is wound with more turns than the other. For example, if the primary winding has 200 turns and the secondary 1000 turns, the voltage available at the secondary terminals will be $\frac{1000}{200}$, or 5 times as great as the voltage applied to the primary winding. This ratio of turns (N_2) in the secondary to the number of turns (N_1) in the primary is called the turns or transformation ratio (r) and it is expressed by the equation.

$$r = \frac{N_2}{N_1} = \frac{E_2}{E_1}$$

where E_1 and E_2 are the respective voltages of the two windings.

When the transformation ratio is such that the transformer delivers a higher secondary voltage than the primary voltage it is said to be of the "step-up" type. Conversely, a "step-down" transformer is one which lowers the secondary voltage. The circuit arrangements for both types are also shown in Fig. 3.10.

Construction of Voltage Transformers. The core of a voltage transformer is laminated and conventionally is built up of suitably shaped thin stampings, about 0·012 in. thick on average, of silicon-iron or nickel-iron. These materials have the characteristics of fairly high resistivity and low hysteresis; therefore, in the laminated form, the effects of both eddy currents and hysteresis are reduced to a minimum. Two different forms of construction are in common use.

In one the laminations are L-shaped and are assembled to provide a single magnetic circuit; in this form it is used for the transformation of single-phase a.c. The second, known as the shell type, can be used for either single-phase or three-phase transformation and is one in which half the laminations are U-shaped and the remainder are T-shaped, all of them being assembled to give a magnetic circuit with two paths. In both forms of construction the joints are staggered in order to minimize the magnetic leakage at the joints. The laminations are held together by core clamps.

In some designs the cores are formed of strips which are wound rather like a clock spring and bonded together. The cores are then cut into two C-shaped parts to allow the pre-wound coils to be fitted. The mating surfaces of the two parts are often ground to give a very small effective gap which helps to minimize the excitation current. After assembly of the windings the core parts are clamped together by a steel band around the outside of the core.

Transformer windings are of enamelled copper wire or strip, and are normally wound on the core one upon the other, to obtain maximum mutual inductive effect, and are well insulated from each other. An exception to this normal arrangement is in a variant known as an auto-transformer, in which the windings are in series and on a core made up of L-shaped laminations. Part of both primary and secondary windings are wound on each side of the core. On a shell-type transformer both windings are wound on the centre limb for single-phase operation, and for three-phase operation they are wound on each limb. Alternative tappings are generally provided on both windings of a transformer for different input and output voltages, while in some types a number of different secondary windings provide simultaneous outputs at different voltages.

Circuit Connections. Voltage transformers are connected so that the primary windings are in parallel with the supply voltage; the primary windings of current transformers are connected in series. A single-phase transformer as the name suggests is for the transformation of voltage from a single-phase supply or from any one phase of a three-phase supply. Transformation of three-phase a.c. can be carried out by means of three separate single-phase transformers, or by a single three-phase transformer. Transformers for three-phase circuits can be connected in one of several combinations of the star and delta connections (see also Chapter 2), depending on the requirements for the transformer. The arrangements are illustrated in Fig. 3.11.

When the star connection is used in three-phase transformers for the operation of three-phase consumer equipment, the transformer may be connected as a three-phase system (Fig. 3.11(a)). If single-phase loads have to be powered from a three-phase supply it is sometimes difficult to keep them balanced, it is therefore essential to provide a fourth or neutral wire so that connections of the loads may be made between this wire and any one of the three-phase lines (Fig. 3.11(b)).

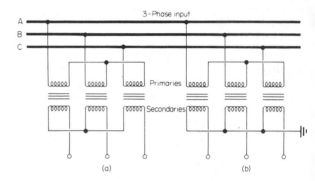

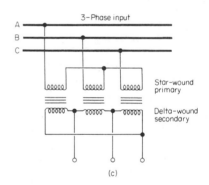

Fig 3.11
Circuit connections for three-phase transformers
(a) Star connection three-wire
(b) Star connection four-wire
(c) Star and Delta connection

The interconnection of neutral points of two star windings is sometimes undesirable because this provides an external path for the flow of certain harmonic currents which can lead to interference with radio communications equipment. This is normally overcome by connecting one of the two transformer windings in delta, for example, if the transformer supplies an unbalanced load, the primary winding is in star and the secondary is in delta as shown in Fig. 3.11(c).

CURRENT TRANSFORMERS

Current transformers are used in many a.c. generator regulation and protection systems and also in conjunction with a.c. ammeters. These transformers have an input/output current relationship which is inversely proportional to the turns ratio of the primary and secondary windings. A typical unit is shown in Fig. 3.12. It is designed with only a secondary winding on a toroidal strip-wound core of silicon-iron. The assembly together with the metal base is encapsulated in a resin compound moulding. The polarity of the transformer is indicated by the markings H1 on the side facing the generator and H2 on the side facing the load.

The primary winding is constituted by passing a main cable of the power system, through the core aperture. The cable is wound with a single turn if it carries high currents, and with two or three turns if it carries low currents. The operating principle is the same as that of a conventional transformer.

In some aircraft generating systems, a number of current transformers are combined into single package assemblies to provide a means of centralizing equipment location. One such assembly is illustrated in Fig. 3.13. It consists of seven transformers which are supplied with primary voltage via the three feeder terminals and by insulated busbars passing through the cores of the transformers which are arranged in three sets. The busbars terminate in the flexible insulated straps. Secondary leads from the various

transformers are brought out through a common connector.

Contrary to the practice adopted for voltage transformers, whenever the secondary windings of current transformers are disconnected from their load circuits, terminals must be short-circuited together. If this is not done, a dangerous voltage may develop which may be harmful to anyone accidentally touching the terminals, or may even cause an electrical breakdown between the windings.

AUTO-TRANSFORMERS

In circuit applications normally requiring only a small step-up or step-down of voltage, a special variant of transformer design is employed and this is known as an auto-transformer. Its circuit arrangement is shown in Fig. 3.14 and from this it will be noted that its most notable feature is that it consists of a single winding tapped to form primary and secondary parts. In the example illustrated the tappings provide a stepped-up voltage output, since the number of primary turns is less than that of the secondary turns.

When a voltage is applied to the primary terminals current will flow through the portion of the winding spanned by these terminals. The magnetic flux due to this current will flow through the core and will therefore, link with the whole of the winding. Those turns between the primary terminals act in the same way as the primary winding of a conventional transformer, and so they produce a self-induction voltage in opposition to the applied voltage. The voltage induced in the remaining turns of the winding will be additive, thereby giving a secondary output voltage

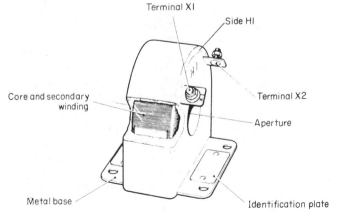

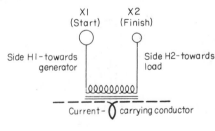

Fig 3.12
Current transformer

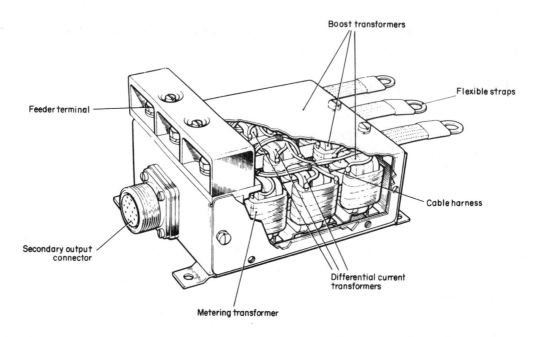

Fig 3.13
Current transformer package

greater than the applied voltage. When a load circuit is connected to the secondary terminals, a current due to the induced voltage will flow through the whole winding and will be in opposition to the primary current from the input terminals. Since the turns between the primary terminals are common to input and output circuits alike they carry the difference

between the induced current and primary current, and they may therefore be wound with smaller gauge wire than the remainder of the winding.

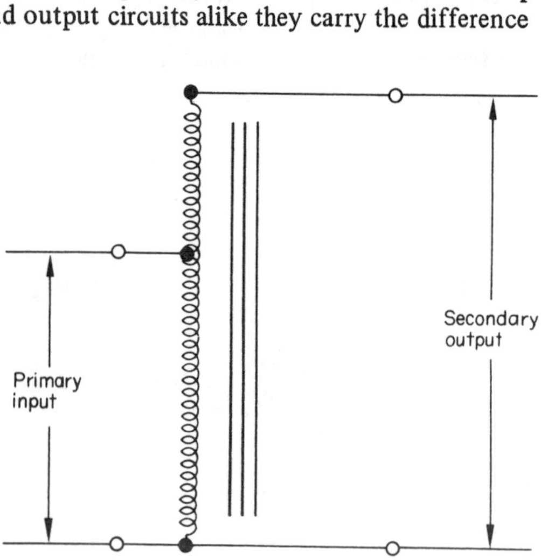

Auto-transformers may also be designed for use in consumer circuits requiring three-phase voltage at varying levels. The circuit arrangement of a typical step-up transformer applied to a windshield anti-icing circuit is shown in Fig. 3.15. The three windings are star-connected and are supplied with the "primary" voltage of 208 volts from the alternator system. The secondary tappings are so arranged that up to four output voltage levels may be utilized.

TRANSFORMER RATINGS

Transformers are usually rated in volt-amperes or kilovolt-amperes. The difference between the output terminal voltages at full-load and no-load, with a constant input voltage, is called the regulation of the transformer. As in the case of an a.c. generator, regulation is expressed as a percentage of the full-load voltage, and depends not only on actual losses (e.g. hysteresis, eddy current and magnetic leakage) but also on the power factor of the load. Thus, an inductive load, i.e. one having a lagging power factor, will give rise to a high percentage regulation, while with a capacitive load, i.e. one having a leading power

Fig 3.14
Circuit arrangement of an auto-transformer

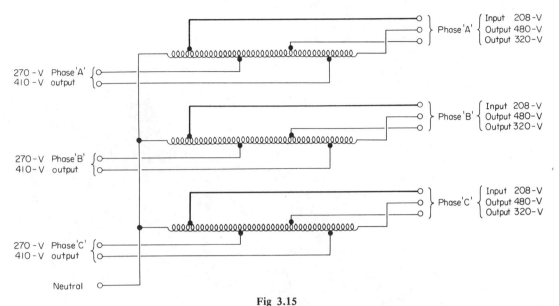

Fig 3.15

Tappings of a typical three-phase auto-transformer

factor, the regulation may be a negative quality giving a higher output voltage on full-load than on no-load.

Changes in power supply frequency, or the connection of a transformer to a supply whose frequency differs from that for which the transformer was designed, has a noticeable effect on its operation. This is due to the fact that the resistance of primary windings are so low that they may be considered to be a purely inductive circuit. If, for example, the frequency is reduced at a constant value of voltage, then the current will rise. The increased current will, in turn, bring the transformer core nearer to magnetic saturation and this decreases the effective value of inductance leading to still larger current. Thus, if a transformer is used at a frequency lower than that for which it was designed, there is a risk of excessive heat generation in the primary winding and subsequent burn out. On the other hand, a transformer designed for low frequency can be used with higher frequencies, since in this case the primary current will be reduced.

TRANSFORMER-RECTIFIER UNITS

Transformer-rectifier units (T.R.U.'s) are combinations of static transformers and rectifiers, and are utilized in some a.c. systems as secondary supply units, and also as the main conversion units in aircraft having rectified a.c. power systems.

Fig. 3.16 illustrates a T.R.U. designed to operate on a regulated three-phase input of 200 volts at a frequency of 400 Hz and to provide a continuous d.c. output of 110 A at approximately 26 volts. The circuit is shown schematically in Fig. 3.17. The unit consists of a transformer and two three-phase bridge rectifier assemblies mounted in separate sections of the casing. The transformer has a conventional star-wound primary winding and secondary windings wound in star and delta. Each secondary winding is connected to individual bridge rectifier assemblies made up of six silicon diodes, and connected in parallel. An ammeter shunt (dropping 50 mV at 100 A) is connected in the output side of the rectifiers to enable current taken from the main d.c. output terminals to be measured at ammeter auxiliary terminals. These terminals, together with all others associated with input and output circuits, are grouped on a panel at one end of the unit. Cooling of the unit is by natural convection through gauze-covered ventilation panels and in order to give warning of overheating conditions, thermal switches are provided at the transformer and rectifier assemblies, and are connected to independent warning lights. The switches are supplied with d.c. from an external source (normally one of the busbars) and their contacts close when temperature conditions at their respective locations rise to approximately 150°C and 200°C.

64

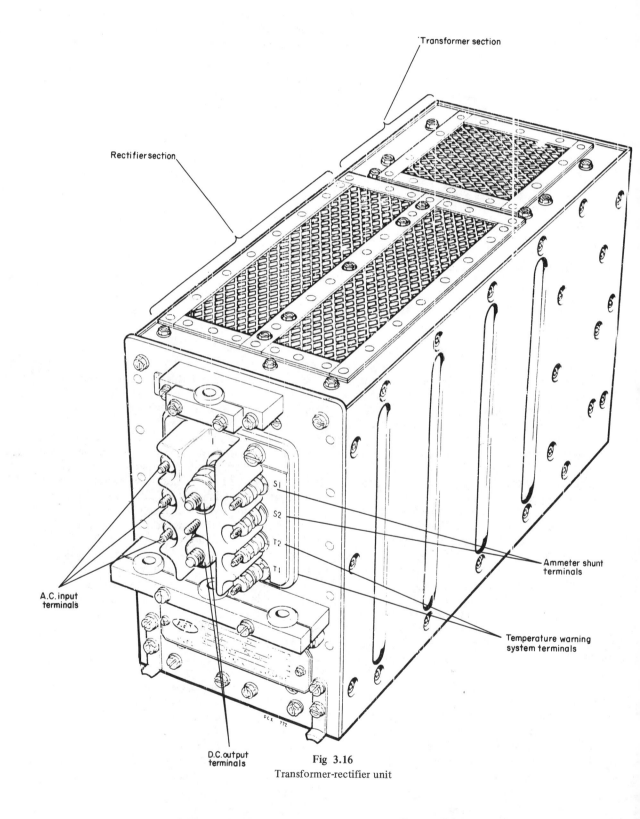

Transformer section

Rectifier section

S1

S2

T2

T1

Ammeter shunt
terminals

A.C. input
terminals

Temperature warning
system terminals

D.C. output
terminals

Fig 3.16
Transformer-rectifier unit

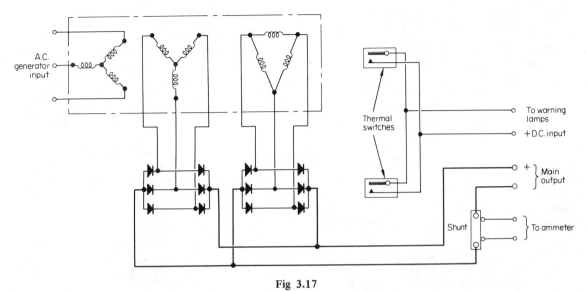

Fig. 3.17
Schematic circuit of a transformer-rectifier unit

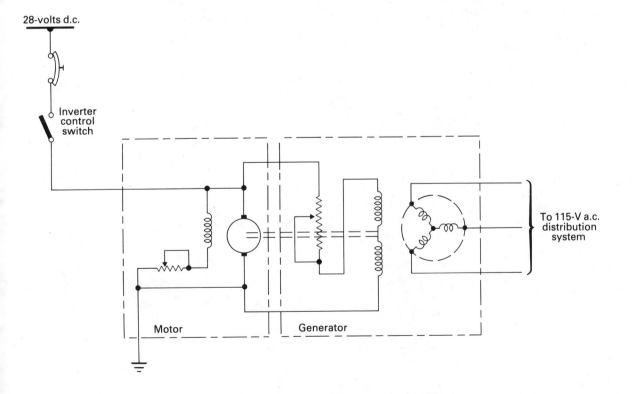

Fig 3.18
Rotary inverter operation

Rotary Converting Equipment

The most widely-known device under this heading is the inverter designed to produce either 26 volts or 115 volts 400 Hz a.c. depending on the secondary a.c. power requirements of an aircraft's electrical system. Although now largely superseded by inverters of the solid-state circuit or static type, rotary inverters are still utilized in a number of the smaller types of aircraft.

A rotary inverter consists of a d.c. motor driving an a.c. generator, and since many of the systems which are to be operated from it are dependent on constant voltage and frequency, the a.c. supply must be regulated accordingly. The methods of regulation may vary, but we may consider the commonly adopted method shown in Fig. 3.18.

When the inverter is switched on, d.c. is supplied to the motor armature and shunt field winding, and also to the excitation field winding of the generator. Thus, the motor starts driving the generator which will produce a three-phase a.c. output at 115 volts. In order to control the voltage at this

level, the d.c. supply is passed through a resistor in series with the generator field. This resistor is preset to give the required excitation current at the regulated d.c. system voltage level. Since the frequency of the generator output is dependent on speed, then a preset resistor is also connected in series with the motor shunt field to provide sufficient excitation current to run the motor and generator at the speed necessary to produce a 400 Hz output.

Figure 3.19 illustrates a sectional view and circuit arrangement of another type of rotary inverter, and

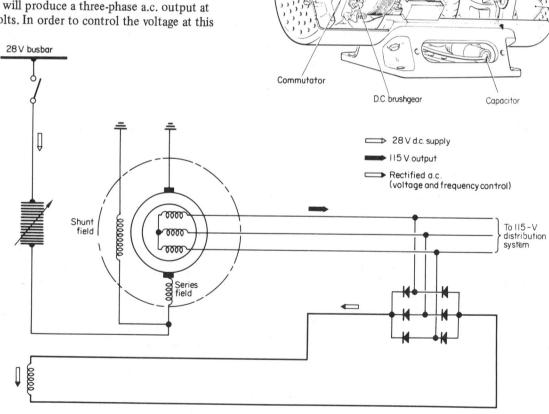

Fig 3.19
Rotary inverter (carbon-pile regulation)

although it is only to be found on some older types of aircraft, it is an interesting example of variation in application of principles.

The motor and generator share a common armature and field system, and control of voltage and frequency is based on the carbon pile regulator principle (see page 12).

The d.c. section of the machine is of the four-pole compound-wound type, the d.c. being supplied to the armature winding, series and shunt-field windings. The a.c. section corresponds to a star-wound generator, the winding being located in the slots of the armature and beneath the d.c. winding. The a.c. winding is connected to a triple slip ring and brushgear assembly at the opposite end to the commutator. Thus, when the inverter is in operation, a three-phase output is induced in a rotating winding and not a fixed stator winding as in the case of a conventional a.c. generator.

The a.c. output is rectified and supplied to the voltage coil of the regulator which varies the pile resistance in the usual manner, this, in turn, varying the current flow through the common field system to keep both the voltage and frequency of the a.c. output within limits.

STATIC INVERTERS

These inverters perform the same conversion function as the rotary machines described earlier, but by means of solid-state or static circuit principles. They are employed in a number of types of aircraft in some cases as a normal source of a.c. power, but more usually to provide only emergency a.c. power to certain essential systems when a failure of the normal 115-volts source has occurred. The function of an inverter used for the conversion of battery supply to single-phase 115-volts a.c. is shown in the block diagram of Fig. 3.20.

The d.c. is supplied to transistorized circuits of a filter network, a pulse shaper, a constant current generator, power driver stage and the output stage. After any variations in the input have been filtered or smoothed out, d.c. is supplied to a square-wave generator which provides first-stage conversion of the d.c. into square-wave form a.c. and also establishes the required operating frequency of 400 Hz. This output is then supplied to a pulse shaper circuit which controls the pulse width of the signal and changes its wave form before it is passed on to the power driver stage. It will be noted from the diagram that the d.c. required for pulse shaper operation is supplied via a turn-on

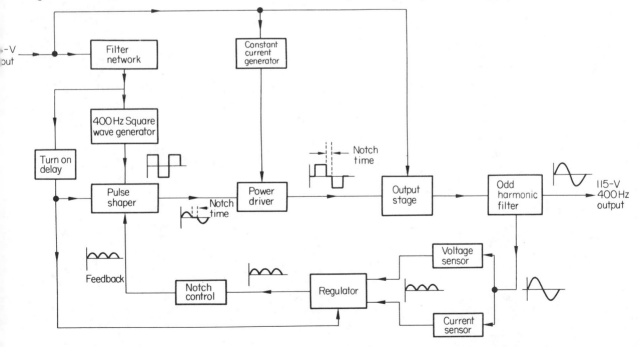

Fig 3.20
Static inverter principle

delay circuit. The reason for this is to cause the pulse shaper to delay its output to the power driver stage until the voltage has stabilized. The power driver supplies a pulse-width modulated symmetrical output to control the output stage, the signal having a square-wave form. The power driver also shorts itself out each time the voltage falls to zero, i.e. during "notch time".

The output stage also produces a square-wave output but of variable pulse width. This output is finally fed to a filter circuit which reduces the total odd harmonics to produce a sine wave output at the volt-age and frequency required for operating the systems connected to the inverter.

As in the case of other types of generators, the output of a static inverter must also be maintained within certain limits. In the example illustrated, this is done by means of a voltage sensor and a current sensor, both of which produce a rectified a.c. feed-back signal which controls the "notch time" of the pulse shaper output through the medium of a regula-tor circuit and a notch control circuit.

CHAPTER FOUR

External and Auxiliary Power Supplies

Electrical power is required for the starting of engines, operation of certain services during "turn-round" servicing periods at airports, e.g. lighting, and for the testing of electrical systems during routine maintenance checks. The batteries of an aircraft are, of course, a means of supplying the necessary power, and although capable of effecting engine starts their capacity does not permit widescale use on the ground and as we have already learned from Chapter 1, they are restricted to the supply of power under emergency conditions. It is necessary, therefore, to incorporate a separate circuit through which power from an external ground power unit (see Fig. 4.1) may be connected to the aircraft's distribution busbar system. In its simplest form, an external power supply system consists of a connector located in the aircraft at a conveniently accessible point (at the side of a fuselage for example) and a switch for completing the circuit between the ground power unit and the busbar system.

In addition to the external power supply system, some types of aircraft carry separate batteries which

can supply the ground services in the event that a ground power unit is not available in order to conserve the main batteries for engine starting.

In the majority of large public transport aircraft, complete independence of ground power units is obtained by special auxiliary power units installed within the aircraft.

D.C. Systems

A basic system for the supply of d.c. is shown in Fig. 4.2, and from this it will also be noted how, in addition to the external power supply, the battery may be connected to the main busbar by selecting the "flight" position of the switch. As the name suggests

Fig 4.1
Ground power unit

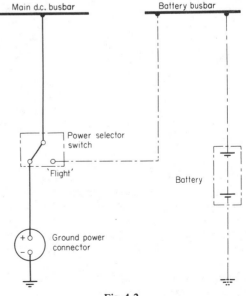

Fig. 4.2
Basic external power supply system

this is the position to which the switch is selected when the aircraft is in flight since under this condition the generator system supplies the main busbar and the battery is constantly supplied with charging current.

The external power connector symbol shown in the diagram represents a twin-socket type of unit which although of an obsolete type is worth noting because it established certain aspects which are basic in the design of present-day connectors or receptacles as they are also called, namely the dimensioning of pins and sockets, and the method of protecting them. The pins were of different diameters to prevent a reverse polarity condition, and the cover of the unit had to be rotated to expose the sockets.

An example of a current type of unit is shown in Fig. 4.3. It consists of two positive pins and one negative pin; one of the positive pins is shorter and of smaller diameter than the remaining pins. The pins are enclosed by a protective shroud, and the complete unit is normally fitted in a recessed housing located at the appropriate part of the airframe structure. Access to the plug from outside the aircraft, is via a hinged flap provided with quick-release fasteners.

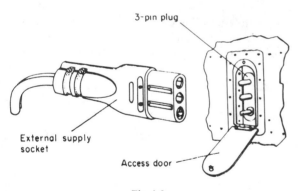

Fig 4.3
External power supply connection

The circuit of a three-pin receptacle system is illustrated in Fig. 4.4, and from this it will be noted that the short positive pin is connected in the coil circuit of the external power relay. The reason for this is that in the event of the external supply socket being withdrawn with the circuit "live", the external power relay will de-energize before the main pins are disengaged from the socket. This ensures that breaking of the supply takes place at the heavy-duty contacts of the relay thus preventing arcing at the main pins.

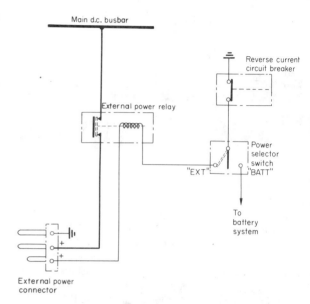

Fig 4.4
Three-pin receptacle system

In some aircraft d.c. power is distributed from a multiple busbar system and it is necessary for certain services connected to each of the busbars to be operated when the aircraft is on the ground. This requires a more sophisticated arrangement of the external power supply system and the circuit of one such arrangement is shown in Fig. 4.5. In addition to the external supply relay or contactor, contactors for "tying" busbars together are provided, together with magnetic indicators to indicate that all connections are made.

When the external ground power unit is connected to the aircraft and the master switch is selected "on", it energizes the external supply contactor, thus closing its auxiliary and main sets of contacts. One set of auxiliary contacts complete a circuit to a magnetic indicator which then indicates that the external supply is connected and on ("C" in Fig. 4.5), a second set complete circuits to the coils of No. 1 and No. 3 bus-tie contactors while a third and main heavy-duty set connect the supply direct to the "vital" and No. 2. busbars. When both bus-tie contactors are energized their main contacts connect the supply from the external supply contactor to their respective busbars. Indication that both busbars are also "tied" to the ground power supply is provided by magnetic indicators "A" and "B" which are energized from the vital busbar via the auxiliary contacts of the contactor.

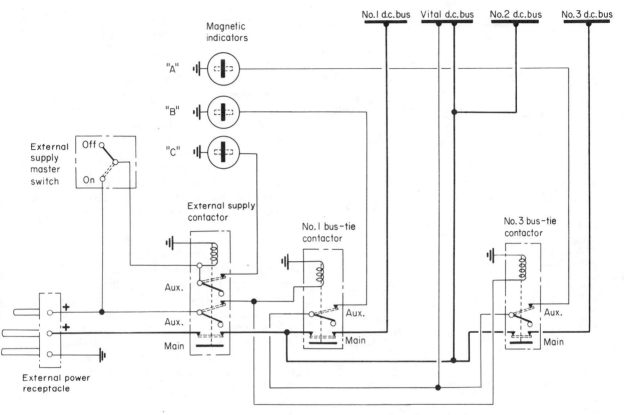

Fig 4.5
Schematic of an external power supply – multiple d.c.
busbar system

In some aircraft, and as an example we may consider the Boeing 737, a separate external power connector is installed for starting an auxiliary power unit in the event that the aircraft's battery is inoperative. The circuit arrangement is shown in Fig. 4.6.

The receptacle is located adjacent to the battery together with two circuit breakers indicated as "A" and "B" in the diagram. The positive pin of the receptacle is coupled directly to the battery busbar via circuit breaker "A", and forms a parallel circuit with the battery. Before external power is applied, circuit breaker "B" must be tripped in order to prevent damage to the battery charger.

A.C. Systems

In aircraft which from the point of view of electrical power are principally of the "a.c. type", then it is

essential for the external supply system of the installation to include a section through which an external source of a.c. power may be supplied. The circuit arrangements for the appropriate systems vary between aircraft types but in order to gain some understanding of the circuit requirements and operation generally we may consider the circuit shown in Fig. 4.7.

When external power is coupled to the receptacle a three-phase supply is fed to the main contacts of the external power breaker, to an external power transformer/rectifier unit (T.R.U.) and to a phase sequence protection unit. The T.R.U. provides a 28 volt d.c. feedback supply to a hold-in circuit of the ground power unit. If the phase sequence is correct the protection unit completes a circuit to the control relay coil, thus energizing it. A single-phase supply is also fed to an amber light which comes on to indicate that external power is coupled, and to a voltmeter and

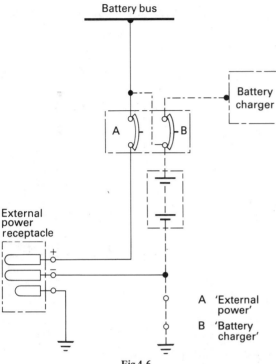

Fig 4.6
Separate external d.c. supply for A.P.U. starting

frequency meter via a selector switch.

The circuit is controlled by an external power switch connected to a busbar supplied with 28 volts d.c. from the aircraft battery system. When the switch is set to the "close" position current flows across the main contacts of the energized control relay, to the "close" coil of the external power breaker, thus energizing it to connect the external supply to the three-phase a.c. main busbar. The external power supply is disconnected by selecting the "trip" position on the external power switch. This action connects a d.c. supply to the trip coil of the external power breaker, thus releasing its main and auxiliary contacts and isolating the external power from the a.c. main busbar.

Figure 4.8 illustrates an external a.c. power receptacle and control panel arrangement generally representative of that adopted in large public transport aircraft. The receptacle is of the six-prong type; three of the large prongs are for the corresponding a.c. power phases, and a fourth large prong for the ground connection between the aircraft structure and external power unit. The two small shorter prongs connect d.c. power for the operation of interlocking relays which connect the external a.c. power to the aircraft.

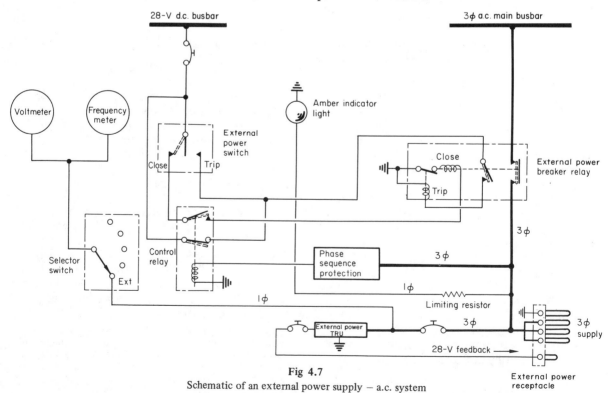

Fig 4.7
Schematic of an external power supply — a.c. system

The control panel contains three single-phase a.c. circuit breakers, and three more breakers which protect relay control and indicating light circuits within the aircraft's external power supply circuit. Indicator lights, interphone jack plug sockets, and pilot's call button switch are also contained on the panel.

The white indicator light is only illuminated whenever external a.c. power is connected but is not supplying power to any a.c. load busbar on the aircraft. The blue light is illuminated whenever a.c. power is being supplied to the load busbars.

The pilot's call button switch and interphone jack plug sockets provide for communication between ground crew and flight crew.

AUXILIARY POWER UNITS

Many of today's aircraft are designed so that if necessary, they may be independent of ground support equipment. This is achieved by the incorporation of an auxiliary power unit (A.P.U.) in the tail section which, after being started by the aircraft's battery system, provides power for engine starting, ground air conditioning and other electrical services. In some installations, the A.P.U. is also used for supplying power in flight in the event of an engine-driven generator failure (see p. 16) and for supplementing the delivery of air to the cabin during take-off and climb.

In general, an A.P.U. consists of a small gas turbine engine, a bleed-air control and supply system, and an accessory gearbox. The gas turbine comprises a two-stage centrifugal compressor connected to a single-stage turbine. The bleed-air control and supply system automatically regulates the amount of air bleed from the compressor for delivery to the cabin air conditioning system. In addition to those accessories essential for engine operation, e.g. fuel pump control unit and oil pumps, the accessory gearbox drives a generator which is of the same type as those driven by the main engines, and having the same type of control and protection unit.

A motor for starting the A.P.U. is also secured to the gearbox and is operated by the aircraft battery system or, when available, from a ground power unit. In some types of A.P.U. the functions of engine starting and power generation are combined in a starter/generator unit. In order to record the hours run, an hour meter is automatically driven by an A.P.U.

An external view of a typical unit and a typical installation, are shown in Figs. 4.9 and 4.10 respectively.

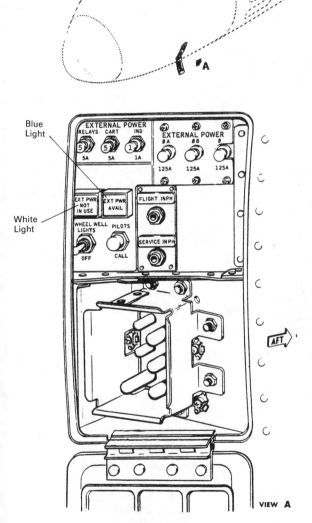

Fig 4.8
External a.c. power receptacle and control panel

74

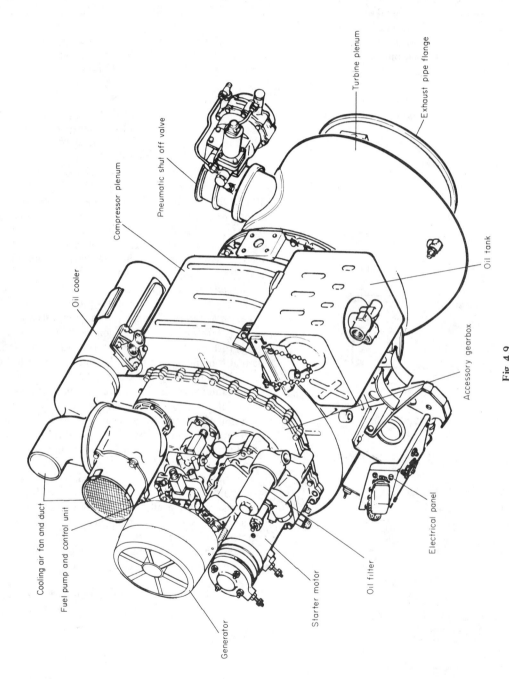

Cooling air fan and duct

Fuel pump and control unit

Oil cooler

Compressor plenum

Pneumatic shut off valve

Turbine plenum

Exhaust pipe flange

Oil tank

Accessory gearbox

Electrical panel

Oil filter

Starter motor

Generator

Fig 4.9
Auxiliary power unit

Fig 4.10
A.P.U. installation

1. Cooling duct
2. Generator
3. Immersion thermocouple switch
4. Oil cooler
5. Hour meter

6. Low oil-pressure switches
7. Fuel filter
8. Oil tank
9. Combustion chamber
10. Ignition exciter

CHAPTER FIVE

Power Distribution

In order for the power available at the appropriate generating sources, to be made available at the inputs of the power-consuming equipment and systems then clearly, some organized form of distribution throughout an aircraft is essential. The precise manner in which this is arranged is governed principally by the type of aircraft and its electrical system, number of consumers and location of consumer components. For example, in a small light aircraft, electrical power requirements may be limited to a few consumer services and components situated within a small area, and the power may be distributed via only a few yards of cable, some terminal blocks, circuit breakers or fuses. In a large multijet transport aircraft on the other hand, literally miles of cable are involved, together with multiple load distribution busbars, protection networks, junction boxes and control panels.

BUSBARS

In most types of aircraft, the output from the generating sources is coupled to one or more low impedance conductors referred to as busbars. These are usually situated in junction boxes or distribution panels located at central points within the aircraft, and they provide a convenient means for connecting positive supplies to the various consumer circuits; in other words, they perform a "carry-all" function. Busbars vary in form dependent on the methods to be adopted in meeting the electrical power requirements of a particular aircraft type. In a very simple system a busbar can take the form of a strip of interlinked terminals while in the more complex systems main busbars are thick metal (usually copper) strips or rods to which input and output supply connections can be made. The strips or rods are insulated from the main structure and are normally provided with some form of protective covering. Flat, flexible strips of braided

copper wire are also used in some aircraft and serve as subsidiary busbars.

Busbar Systems. The function of a distribution system is primarily a simple one, but it is complicated by having to meet additional requirements which concern a power source, or a power consumer system operating either separately or collectively, under abnormal conditions. The requirements and abnormal conditions, may be considered in relation to three main areas, which may be summarized as follows:

1. Power-consuming equipment must not be deprived of power in the event of power source failures unless the total power demand exceeds the available supply.
2. Faults on the distribution system (e.g. fault currents, grounding or earthing at a busbar) should have the minimum effect on system functioning, and should constitute minimum possible fire risk.
3. Power-consuming equipment faults must not endanger the supply of power to other equipment.

These requirements are met in a combined manner by paralleling generators where appropriate, by providing adequate circuit protection devices, and by arranging for faulted generators to be isolated from the distribution system. The operating fundamentals of these methods are described elsewhere in this book, but the method with which this Chapter is concerned is the additional one of arranging busbars and distribution circuits so that they may be fed from different power sources.

In adopting this arrangement it is usual to categorize all consumer services into their order of importance and, in general, they fall into three groups: vital, essential and non-essential.

Vital services are those which would be required after an emergency wheels-up landing, e.g. emergency lighting and crash switch operation of fire extinguishers. These services are connected directly to the battery.

Essential services are those required to ensure safe flight in an in-flight emergency situation. They are connected to d.c. and a.c. busbars, as appropriate, and in such a way that they can always be supplied from a generator or from batteries.

Non-essential services are those which can be isolated in an in-flight emergency for load shedding purposes, and are connected to d.c. and a.c. busbars, as appropriate, supplied from a generator.

Figure 5.1 illustrates in much simplified form, the

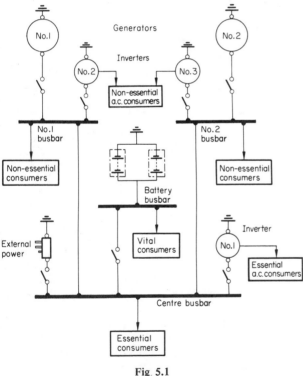

Fig. 5.1
Busbar system

principle of dividing categorized consumer services between individual busbars. In this example, the power distribution system is one in which the power

supplies are 28-volts d.c. from engine-driven generators operating in parallel, 115-volts 400 Hz a.c. from rotary inverters, and 28-volts d.c. from batteries. Each generator has its own busbar to which are connected the non-essential consumer services. Both busbars are in turn connected to a single busbar which supplies power to the essential services. Thus, with both generators operating, all consumers requiring d.c. power are supplied. The essential services busbar is also connected to the battery busbar thereby ensuring that the batteries are maintained in the charged condition. In the event that one generator should fail it is automatically isolated from its respective busbar and all busbar loads are then taken over by the operative generator. Should both generators fail however, non-essential consumers can no longer be supplied, but the batteries will automatically supply power to the essential services and keep them operating for a pre-determined period calculated on the basis of consumer load requirements and battery state of charge.

For the particular system represented by Fig. 5.1, the d.c. supplies for driving the inverters are taken from busbars appropriate to the importance of the a.c. operated consumers. Thus, essential a.c. consumers are operated by No. 1 inverter and so it is driven by d.c. from the essential services busbar. No. 2 and No. 3 inverters supply a.c. to non-essential services and so they are powered by d.c. from the No. 1 and No. 2 busbars.

Figure 5.2 illustrates a split busbar method of power distribution, and is based on an aircraft utilizing non-paralleled constant-frequency a.c. as the primary power source and d.c. via transformer-rectifier units (T.R.U.'s).

The generators supply three-phase power through separate channels, to the two main busbars and these, in turn, supply the non-essential consumer loads and T.R.U.'s. The essential a.c. loads are supplied from the essential busbar which under normal operating conditions is connected via a changeover relay to the No. 1 main busbar. The main busbars are normally isolated from each other i.e., the generators are not paralleled, but if the supply from either of the generators fails, the busbars are automatically inter-connected by the energizing of the "bus-tie" breaker and serve as one, thereby maintaining supplies to all a.c. consumers and both T.R.U.'s. If, for any reason, the power supplied from both generators should fail the non-essential services will be isolated and the changeover relay between No. 1 main busbar, and the essential busbar, will automatically de-energize and connect

78

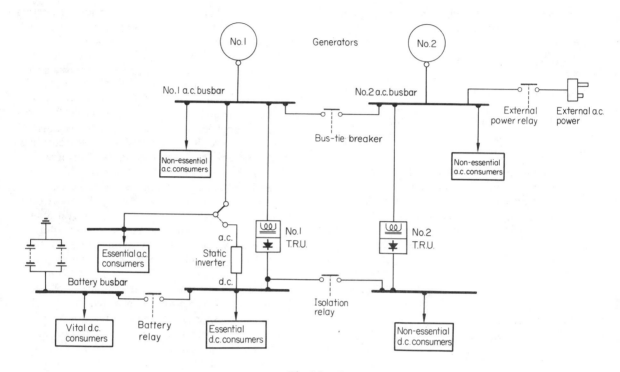

Fig 5.2
Split busbar system (primary a.c. power source)

the essential busbar to an emergency static inverter.

The supply of d.c. is derived from independent T.R.U. and from batteries. The No. 1 T.R.U. supplies essential loads and the No. 2 unit supplies non-essential loads connected to the main d.c. busbar; both busbars are automatically interconnected by an isolation relay. The batteries are directly connected to the battery busbar and this is interconnected with the essential busbar. In the event of both generators failing the main d.c. busbar will become isolated from the essential d.c. busbar which will then be automatically supplied from the batteries to maintain operation of essential d.c. and a.c. consumers.

External power supplies and supplies from an auxiliary power unit can be connected to the whole system in the manner indicated in Fig. 5.2.

Another example of a split busbar system, based on that used in the B737, is shown in Fig. 5.3. The primary power source is non-paralleled 115/200-volt 3-phase a.c. from two 40 kVA generators. A source of a.c. power can be supplied from another 40 kVA generator driven by an auxiliary power unit, and also from an external power unit. Direct current is supplied via three T.R.U.'s.

The four power sources are connected to the busbars by six 3-phase breakers and two transfer relays, which are energized and de-energized according to the switching selections made on the system control panel shown in Fig. 5.4. An interlocking circuit system between breakers and switches is also provided to enable proper sequencing of breaker and overall system operation. A source of power switched onto or entering the system always takes priority and so will automatically disconnect any existing power source. The switches are of the "momentary select" type in that following a selection they are returned

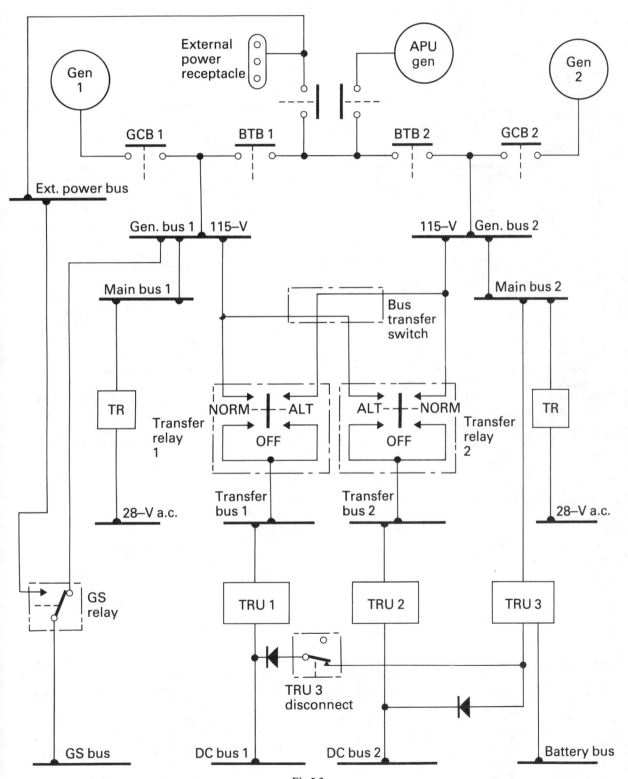

Fig 5.3
Main a.c. and d.c. power distribution system (non-parallel)

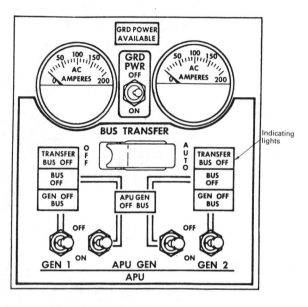

Fig 5.4
Control panel

to a neutral position by spring loading. The bus transfer switch is retained in the "auto" position by a guard cover to provide a path for signals controlling the "normal" and "alternate" positions of the transfer relays. In the "off" position the transfer relays are prevented from being energized to the "alternate" positions so that the two main generating systems are completely isolated from each other.

The indicating lights on the control panel are illuminated as follows:

Ground Power Available (blue)	— when external power is plugged into the aircraft.
Transfer Bus Off (amber)	— when either the normal coil or alternate coil of a transfer relay is de-energized.
Bus Off (amber)	— if both the respective GCB and BTB are open.
Gen Bus Off (blue)	— if the respective GCB is open.
APU Gen Bus Off (blue)	— if APU engine is running and over 95% rev/min, but there is no power from the generator.

The ammeters indicate the load current of both main generators.

When external power is connected to the aircraft and is switched on, the external power contactor closes and energizes both bus-tie breakers (BTB's) to connect power to the whole busbar system. The connection between the generator busbars and transfer busbars is made via the transfer relays which are energized to the "normal" position by the BTB's.

After starting an engine, number 1 for example, and switching on its generator, BTB 1 trips open to allow GCB 1 to close so that all system 1 busbars are supplied from the generator. The number 2 system busbars are still supplied from external power. When number 2 engine has been started and its generator switched on, BTB 2 trips open, GCB 2 closes to connect the generator to the number 2 system busbars, and the external power contactor also trips open.

If it is only necessary for the services connected to the ground service busbar to be operated from external power, this may be effected by leaving the ground power switch on the control panel in the "off" position, and switching on a separate ground service switch. The switch energizes a ground service relay the contacts of which change over a connection from generator bus 1 to the external power busbar.

The APU generator is connected to the entire busbar system via its own three-phase breaker, this, in turn, being energized by two generator switches (see Fig. 5.4). Placing the left or number 1 switch to "on" closes the APU generator breaker and also BTB 1, and with the right or number 2 switch placed to "on" the BTB 2 is closed. As in the case of connecting an external power supply, the transfer relays are energized to the "normal" position by the BTB's.

The normal in-flight configuration of the power distribution system is for each generator to supply its respective busbars through its own breaker, i.e. GCB1 and GCB 2. These breakers are then energized by the generator switches, the interlock circuits keep the BTB's 1 and 2 in the open position, so that the generator systems are always kept entirely separate. GCB 1 and GCB 2 have a set of auxiliary contacts which in the closed position energize transfer relays to their "normal" positions and so provide connections between generators and transfer busbars 1 and 2. As will be noted from the diagram, the transfer busbars supply TRU's 1 and 2, while TRU 3 is supplied direct from the main busbar 2.

In the event of loss of power from one or other

generator, number 1 for example, GCB 1 will open thus isolating the corresponding busbars. When GCB 1 opens, however, another set of auxiliary contacts within the breaker permit a d.c. signal to flow from the control unit of generator 2, via a bus transfer switch, to the "alternate" coil of transfer relay 1. The contacts therefore change over so that power can then be supplied to transfer bus 1 from generator 2 which is still supplying its busbars in the normal way. A similar transfer of power takes place in the event of loss of power from generator 2.

Generator busbar 1 and main busbar 1 which carry non-essential loads, can not be supplied with power from generator 2 under the above power loss conditions. If, however, power to these busbars is required, the APU may be started and its number 1 switch placed momentarily to "on", thereby closing the APU breaker and BTB 1. At the same time, transfer relay 1 contacts would change over from "alternate" to "normal" so that the APU supplies the whole number 1 system. If a loss of power from the number 2 system should then occur, it is not possible to connect it to the APU since its number 2 switch is electrically locked out during in-flight operation.

The three TRU's are connected in such a way that the loss of any one unit will not result in the loss of a d.c. busbar. The relay between TRU 1 and TRU 3 is held closed by supplying d.c. signals from the generator control units via the bus transfer switch in its "auto" position.

A further variation of the split busbar concept, as adopted in the a.c. power generating system of the B747, is simply illustrated in Fig. 5.5. It utilizes a system of interlocking GCB's and BTB's, but in this case various combinations of generator operation are possible.

If the GCB's only are closed, then each generator will only supply its respective load busbar; in other words, they are operated individually and unparalleled. The generators may, however, also be operated in parallel when the BTB's are closed to connect the load busbars to a synchronous busbar. As will be noted from the diagram, this busbar is split into two parts by a split system breaker (SSB) which, in the open position allows the generators to operate in two parallel pairs. Closing of the SSB connects both parts of the synchronous busbar so that all four generators can operate as a fully paralleled system. By means of the interlocking

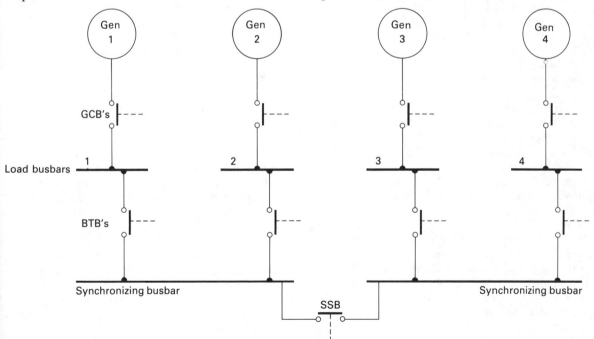

Fig 5.5
Combinations of parallel operation

system between breakers and the manual and automatic sequencing by which they are controlled, any generator can supply power to any load busbar, and any combination of generators can be operated in parallel.

Wires and Cables

Wires and cables constitute the framework of power distribution systems conducting power in its various forms and controlled quantities, between sections contained within consumer equipment (known as "equipment" wires and cables), and also between equipment located in the relevant areas of an aircraft (known as "airframe" wires and cables). The differences between a wire and a cable relate principally to their constructional features (and indirectly to their applications also) and may be understood from the following broad definitions.

A *wire* is a single solid rod or filament of drawn metal enclosed in a suitable insulating material and outer protective covering. Although the term properly refers to the metal conductor, it is generally understood to include the insulation and covering. Specific applications of single wires are to be found in consumer equipment; for example, between the supply connections and the brush gear of a motor, and also between the various components which together make up the stages of an electronic amplifier.

A *cable* is usually made up of a conductor composed of a group of single solid wires stranded together to provide greater flexibility, and enclosed by insulating material and outer protective covering. A cable may be either of the single core type, i.e., with cores stranded together as a single conductor, or of the multicore type having a number of single core cables in a common outer protective covering.

Having highlighted the above definitions, it is interesting to note that with the present lack of international standardization of terminology, they may not be used in the same context. For example, in the U.S. and some other countries, the term "wire" is used as an all-embracing one.

In connection with power distribution systems in their various forms, such terms as "wiring systems", "wiring of components", "circuit wiring" are commonly used. These are of a general nature and apply equally to systems incorporating either wires, cables or both.

TYPES OF WIRES AND CABLES

Wires and cables are designed and manufactured for duties under specific environmental conditions and are selected on this basis. This ensures functioning of distribution and consumer systems, and also helps to minimize risk of fire and structural damage in the event of failure of any kind. Table 5.1 gives details of some commonly used general service wires and cables of U.K. manufacture, while typical constructional features are illustrated in Fig. 5.6.

The names adopted for the various types are derived from contractions of the names of the various insulating materials used. For example, "NYVIN" is derived from "NYlon" and from polyVINyl-chloride (P.V.C.); and "TERSIL" is derived from polyesTER and SILicone. Cables may also be further classified by prefixes and suffixes relating to the number of cores and any additional protective covering. For example, "TRINYVIN" would denote a cable made up of three single Nyvin cables, and if suffixed by "METSHEATH" the name would further denote that the cable is enclosed in a *metal* braided *sheath*.

It will be noted from the Table that only two metals are used for conductors, i.e. copper (which may also be tinned, nickel-plated or silver-plated depending on cable application) and aluminium. Copper has a very low specific resistance and is adopted for all but cables of large cross-sectional areas. An aluminium conductor having the same resistance as a copper conductor, has only two-thirds of the weight but twice the cross-sectional area of the copper conductor. This has an advantage where low-resistance short-term circuits are concerned; for example, in power supply circuits of engine starter motor systems.

The insulation materials used for wires and cables must conform to a number of rigid requirements such as, toughness and flexibility over a fairly wide temperature range, resistance to fuels, lubricants and hydraulic fluids, ease of stripping for terminating, non-flammability and minimum weight. These requirements, which are set out in standard specifications, are met by the materials listed in Table 5.1 and in the selection of the correct cable for a specific duty and environmental condition.

To ensure proper identification of cables, standard specifications also require that cable manufacturers comply with a code and mark outer protective coverings accordingly. Such a coding scheme usually sig-

Table 5.1

Type	Specification British B.S.G.	Specification American MIL-W-	Materials Conductor	Materials Insulation & Covering	Ambient temperature range	Application
NYVIN	177	5086A (Type 2)	Tinned Copper or Aluminium	*P.V.C. Compound Glass braid Nylon	−75°C to +65°C	General services wiring except where ambient temperatures are high and/or extended properties of flexibility are required.
PREN			Tinned Copper or Aluminium	Glass braid Polychloroprene Compound	−75°C to + 50°C	
TERSIL	189	8777B(ASG)	Nickel-plated Copper; or Aluminium	Silicone Rubber Polyester tapes Glass braid Polyester fibre Varnish	−75°C to +150°C	
EFGLAS	192	7129B	Nickel-plated Copper	Glass braid P.T.F.E. +	−75°C to +220°C	In high operating temperatures and in areas where resistance to aircraft fluids necessary. Also where severe flexing under low-temperature conditions is encountered e.g, landing gear shock strut switch circuits.
UNIFIRE – "F"			Nickel-plated Copper	Glass braid P.T.F.E. Asbestos felt impregnated with silicone varnish	Up to 240°C	In circuits required to function during or after a fire.
NYVINMETSHEATH			Tinned Copper or Aluminium	As for NYVIN plus an overall tinned-copper braid over-laid with polyester tape, nylon braid and lacquer	−75°C to +65°C	In areas where screening required
FEPSIL	206		Nickel-plated Copper	Silicone Rubber Glass braid and Varnish F.E.P.**	−75°C to +190°C	

* PolyVinylChloride; + PolyTetraFluoroEthylene; **Fluorinated Ethylene Propylene.

nifies, in sequence, the type of cable, country of origin ("G" for U.K. manufacturers) manufacturer's code letter, year of manufacture also by a letter, and its wire gauge size, thus, NYVIN G-AN 22. A colour code scheme is also adopted particularly as a means of tracing the individual cores of multicore cables to and from their respective terminal points. In such cases it is usual for the insulation of each core to be produced in a different colour and in accordance with the appropriate specification. Another method of coding, and one used for cables in three-phase circuits of some types of aircraft, is the weaving of a coloured trace into the outer covering of each core; thus red − (phase A); yellow − (phase B); blue − (phase C). The code may also be applied to certain single-core cables by using a coloured outer covering.

ROUTING OF WIRES AND CABLES

As noted earlier in this chapter, the quantity of wires and cables required for a distribution system depends on the size and complexity of the systems. However, regardless of quantity, it is important that wires and cables be routed through an aircraft in a manner which, is safe, avoids interference with the reception and transmission of signals by such equipment as radio and compass sytems, and which also permits a systematic approach to their identification, installation and removal, and to circuit testing. Various methods, dependent also on size and complexity, are adopted but in general, they may be grouped under

three principal headings: (i) open loom, (ii) ducted loom, and (iii) conduit.

Open Loom. In this method, wires or cables to be routed to and from consumer equipment in the specific zones of the aircraft, are grouped parallel to each other in a bundle and bound together with waxed cording or p.v.c. strapping. A loom is supported at intervals throughout its run usually by means of clips secured at relevant parts of the aircraft structure. An application of the method to an aircraft junction box is shown in Fig. 5.7.

The composition of a cable loom is dictated by such factors as (i) overall diameter, (ii) temperature conditions, i.e. temperature rise in cables when operating at their maximum current-carrying capacity in varying ambient temperature conditions, (iii) type of current, i.e. whether alternating, direct, heavy-duty or light-duty, (iv) interference resulting from inductive or magnetic effects, (v) type of circuit with which cables are associated; this applies particularly to circuits in the essential category, the cables of which must be safe-guarded against damage in the event of short-circuits developing in adjoining cables.

Magnetic fields exist around cables carrying direct current and where these cables must interconnect equipment in the vicinity of a compass magnetic detector element, it is necessary for the fields to be cancelled out. This is achieved by routing the positive and earth-return cables together and connecting the

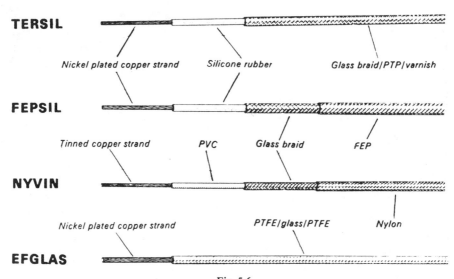

Fig. 5.6
Constructional features of some typical cables

earth-return cable at an earthing point located at a specific safe distance from the magnetic detector element of a compass system.

Ducted Loom. This method is basically the same as that of the open loom except that the bundles are supported in ducts which are routed through the aircraft and secured to the aircraft structure (see Fig. 5.8). Ducts may be of aluminium alloy, resin-impregnated asbestos or moulded fibre-glass-reinforced plastic. In some applications of this method, a main duct containing several channels may be used, each channel supporting a cable loom corresponding to a specific consumer system. For identification purposes, each loom is bound with appropriately coloured waxed cording.

Conduits are generally used for conveying cables in areas where there is the possibility of exposure to oil, hydraulic or other fluids. Depending on the particular application, conduits may take the form of either

plastic, flexible metal or rigid metal sheaths. In cases where shielding against signal interference is necessary the appropriate cables are conveyed by metal conduits in contact with metal structural members to ensure good bonding.

Cable Seals. In pressurized cabin aircraft it is essential for many cables to pass through pressure bulkheads without a "break" in them and without causing leakage

Fig 5.8
Ducted looms

of cabin air. This is accomplished by sealing the necessary apertures with either pressure bungs or pressure-proof plugs and sockets. An example of a pressure bung assembly is shown in Fig. 5.9. It consists of a housing, perforated synthetic rubber bung, anti-friction washer and knurled clamping nuts; the housing is flanged and threaded, having a tapered bore to accept the bung. The holes in the bung vary in size to accommodate cables of various diameters, each hole being sealed by a thin covering of synthetic rubber at the smaller diameter end of the bung. The covering is

Fig 5.7
Open looms

86

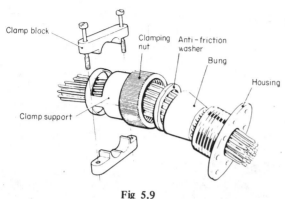

Fig 5.9
Pressure bung assembly

pierced by a special tool when loading the bung with cables.

The cables are a tight fit in the holes of the bung which, when fully loaded and forced into the housing by the clamping nut, is compressed tightly into the housing and around the cables. The anti-friction washer prevents damage to the face of the bung when the clamping nut is turned. On assembly, holes not occupied by cables are plunged with plastic plugs.

In instances where cable "breaks" are required at a pressure bulkhead, the cables at each side of the bulkhead are terminated by specially-sealed plug or socket assemblies of a type similar to those shown in Fig. 5.14 (items 3 and 4).

SPECIAL PURPOSE CABLES

For certain types of electrical systems, cables are required to perform a more specialized function than that of the cables already referred to. Some examples of what are generally termed, special purpose cables, are described in the following paragraphs.

Ignition Cables. These cables are used for the transmission of high tension voltages in both piston engine and turbine engine ignition systems, and are of the single-core stranded type suitably insulated, and screened by metal braided sheathing to prevent interference. The number of cables required for a system corresponds to that of the sparking plugs or igniter plugs as appropriate, and they are generally made up into a complete ignition cable harness. Depending on the type of engine installation, the cables may be enclosed in a metal conduit, which also forms part of a harness, or they may be routed openly. Cables are connected to the relevant system components by

special end fittings comprising either small springs or contact caps secured to the cable conductor, insulation, and a threaded coupling assembly.

Thermocouple Cables. These cables are used for the connection of cylinder head temperature indicators and turbine engine exhaust gas temperature indicators to their respective thermocouple sensing elements. The conducting materials are normally the same as those selected for the sensing element combinations, namely, iron and constantan or copper and constantan for cylinder head thermocouples, chromel (an alloy of chromium and nickel) and alumel (an alloy of aluminium and nickel) for exhaust gas thermocouples.

In the case of cylinder head temperature indicating systems only one thermocouple sensing element is used and the cables between it and a firewall connector are normally asbestos covered. For exhaust gas temperature measurement a number of thermocouples are required to be radially disposed in the gas stream, and it is the usual practice therefore, to arrange the cables in the form of a harness tailored to suit a specific engine installation. The insulating material of the harness cables is either silicone rubber or P.T.F.E. impregnated fibre glass. The cables terminate at an engine or firewall junction box from which cables extend to the indicator. The insulating material of extension cables is normally of the polyvinyl type, since they are subject to lower ambient temperatures than the engine harness. In some applications extension cables are encased in silicone paste within metal-braided flexible conduit.

Co-axial Cables. Co-axial cables contain two or more separate conductors. The innermost conductor may be of the solid, or stranded copper wire type, and may be plain, tinned, silver-plated or even gold-plated in some applications, depending on the degree of conductivity required. The remaining conductors are in the form of tubes, usually of fine wire braid. The insulation is usually of polyethylene or Teflon. Outer coverings or jackets serve to weatherproof the cables and protect them from fluids, mechanical and electrical damage. The materials used for the coverings are manufactured to suit operations under varying environmental conditions.

Co-axial cables have several main advantages. First, they are shielded against electrostatic and magnetic fields; an electrostatic field does not extend beyond the outer conductor and the fields due to current flow in inner and outer conductors cancel each other.

Secondly, since co-axial cables do not radiate, then likewise they will not pick up any energy, or be influenced by other strong fields. The installations in which coaxial cables are most commonly employed are radio, for the connection of antennae, and capacitance type fuel quantity indicating systems for the interconnection of tank units and amplifiers. The construction of a typical coaxial cable and also the sequence adopted for attaching the end fitting are shown in Fig. 5.10. The outer covering is cut back to expose the braided outer conductor (step "A") which

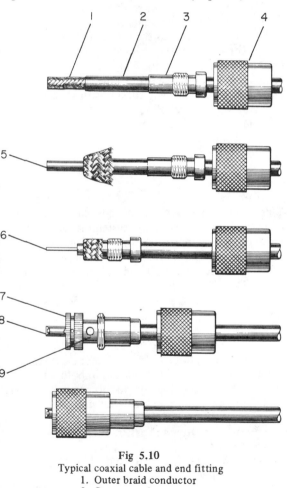

Fig 5.10
Typical coaxial cable and end fitting
1. Outer braid conductor
2. Outer covering
3. Adapter
4. Coupling ring
5. Insulation
6. Inner conductor
7. Plug sub-assembly
8. Contact
9. Solder holes

is then fanned out and folded back over the adapter (steps "B" and "C"). At the same time, the insulation is cut back to expose the inner conductor. The next step (D) is to screw the sub-assembly to the adapter thereby clamping the outer conductor firmly between the two components. Although not applicable to all cables the outer conductor may also be soldered to the sub-assembly through solder holes. The assembly is completed by soldering a contact on to the inner conductor and screwing the coupling ring on to the sub-assembly.

EARTHING OR GROUNDING

In the literal sense, earthing or grounding as it is often termed, refers to the return of current to the conducting mass of the earth, or ground, itself. If considered as a single body, the earth is so large that any transfer of electrons between it and another body fails to produce any perceptible change in its state of electrification. It can therefore be regarded as electrically neutral and as a zero reference point for judging the state of electrification of other bodies. For example, if two charged bodies, A and B, both have positive potentials relative to earth, but the potential of A is more positive than that of B, then the potential of B may be described as negative to that of A by the appropriate amount.

As we have already learned, the positive outputs of aircraft power supplies and the positive input terminals of consumer components are all connected to busbars which are insulated from the aircraft structure. Since in most aircraft the structure is of metal and of sufficient mass to remain electrically neutral, then it too can function as an earth or "negative busbar" and so provide the return path of current. Thus, power supply and consumer circuits can be completed by coupling all negative connections to the structure at various "earth stations", the number and locations of which are predicted in a manner appropriate to the particular type of aircraft. As this results in the bulk of cable required for the circuits being on the positive side only, then such an electrical installation is designated as a "single-wire, or single-pole, earth-return system". For a.c. power supply circuits the airframe also serves as a connection for the neutral point.

The selection of types of connection for earth return cables is based on such important factors as mechanical strength, current to be carried, corrosive effects, and ease with which connections can be made. As a result, they can vary in form; some typical arrangements being a single bolt passing through and secured directly to a structural member, and either a single bolt or a cluster of bolts secured to an earthing plate designed for riveting or bolting to a structural member. In order to ensure good electrical contact and minimum resistance between an earthing bolt or plate and the structure, protective film is removed from the contacting surfaces before assembly. Protection against corrosion is provided by coating the surfaces with an anti-corrosion and solvent resistant compound or, in some cases by interposing an electro-tinned plate and applying compound to the edges of the joint. An example of a cluster arrangement with a corrosion plate is illustrated in Fig. 5.7.

Earth-return cables are connected to earthing bolts by means of crimped ring type connectors, each bolt accommodating cables from several circuits. For some circuits, however, it is necessary to connect cables separately and this applies particularly to those of the sensitive low current-carrying type, e.g. resistance type temperature indicators in which errors can arise from varying earth return currents of other circuits.

In aircraft in which the primary structure is of non-metallic construction, a separate continuous main earth and bonding system is provided. It consists of four or more soft copper strip-type conductors extending the whole length of the fuselage and disposed so that they are not more than six feet apart as measured around the periphery of the fuselage at the position of greatest cross-sectional area. The fuselage earthing strips are connected to further strips which follow the leading and trailing edges from root to tip of each wing and horizontal stabilizer, and also to a strip located on or near the leading edge of the vertical stabilizer. Earthing strips are provided in the trailing edges of the rudder, elevators and ailerons, and are connected to the fuselage and wing systems via the outer hinges of the control surfaces. The strips are arranged to run with as few bends as possible and are connected to each other by means of screwed or riveted joints.

Lightning strike plates, extending round the tips of each wing, horizontal and vertical stabilizers, fuselage nose and tail, are also provided. They consist of copper strips and are mounted on the exterior of the structure.

Connections

In order to complete the linkages between the various units comprising a power distribution system, some appropriate means of connection and disconnection must be provided. The number of connections involved in any one system obviously depends on the type and size of an aircraft and its electrical installation, but the methods of connection with which we are here concerned follow the same basic pattern.

In general, there are two connecting methods adopted and they can be broadly categorized by the frequency with which units must be connected or disconnected. For example, cable connections at junction boxes, terminal blocks, earth stations etc. are of a more permanent nature, but the cable terminations are such that the cables can be readily disconnected when occasion demands. With equipment of a complex nature liable to failure as the result of the failure of any one of a multitude of components, the connections are made by some form of plug and socket thus facilitating rapid replacement of the component. Furthermore, the plug and socket method also facilitates the removal of equipment that has to be inspected and tested at intervals specified in maintenance schedules.

CABLE TERMINATIONS

There are several methods by which cable terminations may be made, but the one most commonly adopted in power distribution systems is the solderless or crimped termination. The soldering method of making connections is also adopted but is more generally confined to the joining of internal circuit connections of the various items of consumer equipment and in some cases, to the connections between single-core cables and plug and socket contacts.

Crimped Terminals. A crimped terminal is one which has been secured to its conductor by compressing it in such a way that the metals of both terminal and conductor merge together to form a homogeneous mass. Some of the advantages of the crimping method are:
1. Fabrication is faster and easier, and uniform operation is assured.
2. Good electrical conductivity and a lower voltage drop is assured.
3. Connections are stronger (approaching that obtained with cold welding); actually as strong as the conductor itself.
4. Shorting due to solder slop and messy flux problems are eliminated.
5. "Wicking" of solder on conductor wires and "dry" joints are eliminated.

6. When properly formed a seal against the ingress of air is provided and a corrosion-proof joint thereby obtained.

A typical terminal (see Fig. 5.11) is comprised of two principal sections; crimping barrel and tongue. For a particular size of conductor the copper or aluminium barrel is designed to fit closely over the barrel end of the conductor so that after pressure has been applied a large number of point contacts are made. The pressure is applied by means of a hand-operated or hydraulically-operated tool (depending on the size of conductor and terminal) fitted with a die, shaped to give a particular cross-sectional form, e.g. hexagonal, diamond or "W". The barrels are insulated by plastic sleeves which extend a short distance over the conductor insulation and provide a certain amount of support for the conductor allowing it to be bent in any direction without fraying of the conductor insulation or breaking of wire strands. In certain types

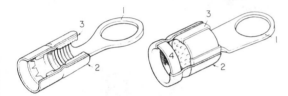

Fig 5.11
Crimped terminals
1. Tongue
2. Insulation sleeve
3. Barrel
4. Stainless steel support
   (large-diameter cable terminals)

of terminal the inside surface of the barrel is serrated so that under the crimping pressure the strands of the conductor "flow" into the serrations to make a connection of high tensile strength. The serrations have the additional function of assisting in the breaking down of the oxide layer that forms on conductor wires during the crimping operation. To facilitate inspection of the crimped joint, the barrel is frequently left open at the tongue end, or in some cases, is provided with an inspection hole through which sufficient insertion of the conductor into the barrel may be visually verified.

The design of the tongue end depends on where and how the terminal is to be attached. The most common forms are the ring type and fork type.

Where a connection between the ends of two cables has to be made, for example, in a cable run from the engine nacelle to the fuselage of an aircraft, a change

from an efglas cable to a nyvin cable may be necessary, a variant of the crimped terminal is used. This variant is known as an in-line connector and consists essentially of two crimping barrels in series, one conductor entering and being crimped at each end. A plastic insulating sleeve is also fitted over the connector and is crimped in position.

A selection of terminals and in-line connectors are shown in Fig. 5.12.

Aluminium Cable Connections. The use of aluminium wire as an electrical conductor for certain systems is due chiefly to the important weight advantage of

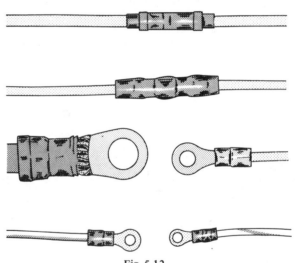

Fig 5.12
Terminals and in-line connectors

this metal over copper. However, in order to acquire satisfactory electrical connections, certain installation techniques are necessary to compensate for two other principal characteristics of aluminium, namely the rapidity with which it oxidizes, and its softness.

The oxide film is formed as soon as aluminium is exposed to the atmosphere and it not only acts as an insulator, but also increases in thickness as heat is generated by the flow of current, still further increasing the electrical resistance and causing corrosion at connecting joints. The method most commonly employed for eliminating the oxide film is the one in which a special zinc granular compound is applied to the exposed ends of the cable and the appropriate terminal. Aluminium terminals are normally of the crimped type and the barrel is filled with compound;

90

in some cases the barrel contains a pre-filled cartridge. When crimping takes place the compound is forced around and between the wire strands of the cable, and penetrates the oxide film to assist in breaking it down. In this manner, clean metal-to-metal contacts are provided and the high electrical resistance of the oxide film is bypassed. Sealing of the terminal/cable joint is also achieved so that the oxide film cannot reform.

In cases where an aluminium cable terminal is to be bolted directly to the aircraft structure, a busbar, or surface of a component, the surfaces are first cleaned and a coating of compound applied. To compensate for the relative softness of aluminium as compared with copper, flat washers with larger diameters than the tongue end of a terminal are used to help distribute the clamping pressure over a wider surface. For reasons of softness also, tightening torques applied to bolted connections are maintained within specific limits.

Plugs and Sockets.   Plugs and sockets (or receptacles) are connecting devices which respectively contain male and female contact assemblies. They may

be fixed or free items, i.e. fixed in a junction box, panel or a consumer component, or free as part of a cable to couple into a fixed item. There are many variations in the design of plugs and sockets governed principally by the distribution circuit requirements, number of conductors to be terminated, and environmental conditions. In general, however, the conventional construction follows the pattern indicated in exploded form in Fig. 5.13. The bodies or shells, are mostly of light alloy or stainless steel finished overall with a cadmium plating; they may be provided with either a male or female thread. Polarizing keys and

Fig 5.14
Typical plugs and sockets
(1, 2) Fixed equipment and panel types
(3, 4) Fixed through-type (bulkhead)
(5) Free type with cable clamp
(6) Fixed type angle fitting
(7) Fixed type rack equipment

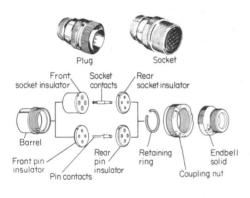

Fig 5.13
Plug and socket construction

keyways are also provided to ensure that plugs and sockets and their corresponding conductors, mate correctly; they also prevent relative movement between their contacts and thereby strain, when the coupling rings are being tightened. The shells of "free" plugs and sockets are extended as necessary by the attachment of outlets or endbells. These provide a means of supporting the cable or cable loom at the point of entry to the plug or socket thereby preventing straining of the conductor, and pin or socket joints, they prevent displacement of the contacts in the softer material insulators, and the ingress of moisture and dirt. In many cases a special cable clamp is also provided (see Fig. 5.14, item 5).

Plug contacts are usually solid round pins, and socket contacts have a resilient section which is arranged to grip the mating pin. The contacts are retained in position by insulators or inserts as they are often called, which are a sliding fit in the shells and are secured by retaining rings and/or nuts. Insulators

may be made from hard plastic, neoprene of varying degrees of hardness, silicone rubber or fluorosilicone rubber depending on the application of a plug and socket, and on the environmental conditions under which they are to be used. Attachment of conductors to pin and socket contacts is done by crimping (see p. 89) a method which has now largely superseded that of soldering. The socket contacts are designed so that their grip on plug pin contacts is not reduced by repeated connection and disconnection.

In most applications, plugs and sockets are secured in the mated condition by means of threaded coupling rings or nuts; in some cases bayonet-lock and push-pull type couplings may also be employed.

Some typical fixed and free type plugs and sockets are illustrated in Fig. 5.14. The rack type unit (item 7) is used principally for the interconnection of radio and other electronic equipment which is normally mounted in special racks or trays. One of the elements, either the plug or the socket, is

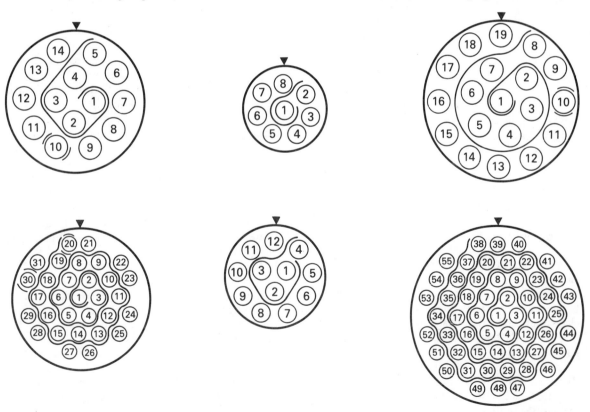

Fig 5.15
Pin/socket sequencing

▼ Polarising key positions

fixed to the back of the equipment and the mating unit is fixed to the rack or tray; electrical connection is made when the equipment is slid into the rack or tray.

In addition to identifying pins and sockets by numbers or letters, it is usual in many types of connectors to signify the numerical or alphabetical sequencing. As shown in Fig. 5.15, this is done by a spiralling "guideline" embossed on the faces of inserts. Every tenth pin or socket cavity is identified with parentheses.

"POTTING"
This is a technique usually applied to plugs and sockets which are to be employed in situations where there is the possibility of water or other liquids passing through the cable entry. It eliminates elaborate cable ferrules, gland nuts, etc, by providing a simple plastic shroud with sufficient height to cover the terminations, and filling the cavity with a special compound which though semi-fluid in its initial condition, rapidly hardens into a rubbery state to form a fairly efficient seal. In addition to sealing it provides reinforcement for the cable connections.

The potting compound consists of a basic material and an alkaline or acid base material (known as an "accelerator") which are thoroughly mixed in the correct proportion to give the desired consistency and hardness of the compound. Once mixed, the compound is injected into a special mould and allowed to set. When the mould is removed, the resilient hemispherically-shaped insulation extends well into the plug or socket, bonding itself to the back of the insulant around the contact and conductor joints and partly out along the conductor insulation.

Electrical Bonding

STATIC CHARGES
During flight, a build-up of electrical energy occurs in the structure of an aircraft, developing in two ways: by precipitation static charges and by charges due to electrostatic induction. Precipitation static charges are built up on the outer surfaces of an aircraft due to frictional contact with rain particles, snow and ice crystals, dust, smoke and other air contamination. As the particles flow over the aircraft negative charges are left behind on the surfaces and positive charges are released to flow into the airstream. In addition, particles of foreign impurities which are themselves charged, make physical contact and transfer these

charges to the surfaces of the aircraft, increasing or decreasing the charged state already present by virtue of the frictional build-up.

Charges of the electrostatic type are those induced into an aircraft when flying into electric fields created by certain types of cloud formation. This condition of charge is the result of the disruption of water particles which increases the strength of a field and builds up such a high voltage that a discharge occurs in the familiar form of lightning. The discharge can take place between oppositely charged pockets in one cloud or a negatively charged section and the top of the cloud, or between a positively charged pocket and earth or ground. A well developed cloud may have several oppositely charged areas, which will produce several electric fields in both the horizontal and vertical planes, where voltages of up to 10,000 volts per centimetre can be achieved. The relative hazard created by these high potentials can be readily appreciated if it is realized that by electrostatic induction, up to 10 million volts with possibly several thousand amperes of current, may be permitted to pass through the aircraft when flying in or near the aforementioned conditions.

Regardless of how an aircraft acquires its static charges the resultant potential difference between it and the atmosphere produces a discharge which tends to adjust the potential of the aircraft to that of the atmosphere. The charge is therefore being dissipated almost as it is being acquired, and by natural means.

One of the hazards, however, is the possibility of discharges occurring within the aircraft as a result of differences between the potentials of the separate parts which go to make up the aircraft, and all the systems necessary for its operation. It is essential, therefore, to incorporate a system which will form a continuous low-resistance link between all parts and in so doing will:

(i) limit the potential difference between all parts.
(ii) eliminate spark discharges and fire risks.
(iii) carry the exceptionally high voltages and currents so that they will discharge to atmosphere at the extremities of the aircraft.
(iv) reduce interference with radio and navigational aid signals.
(v) prevent the possibility of electrical shock hazards to persons contacting equipment and parts of the aircraft.

Such a system is called a *bonding system* and although differing in its principal functions, it will be

93

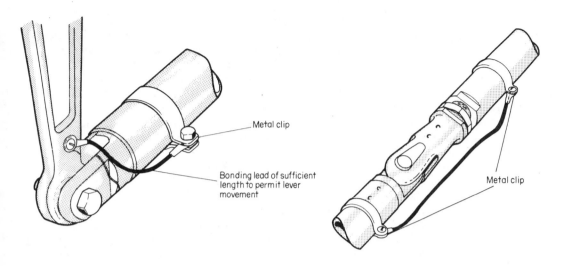

(a) Levers and control rods

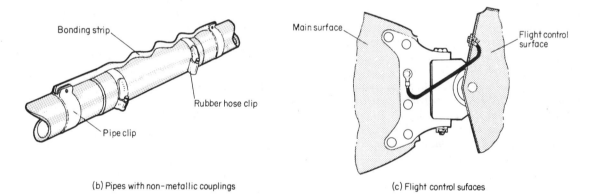

(b) Pipes with non-metallic couplings

(c) Flight control sufaces

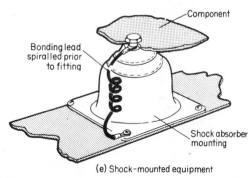

(d) Flexible coupling at bulkheads

(e) Shock-mounted equipment

Fig 5.16
Bonding methods

clear from the fact that electrical continuity is obtained, the requirements of the system overlap those of the earthing system described on p. 87.

The continuous link is formed by metal strip conductors joining fixed metal parts, e.g. pipes joined either side of a non-metallic coupling, and by short-length flexible braid conductors for joining moving parts such as control rods, flight control surfaces, and components mounted on flexible mountings, e.g. instrument panels, mounting racks for electronic equipment. Some typical examples of the method of joining bonding strips or "jumpers" as they are sometimes called, are shown in Fig. 5.16.

In general, bonding is classified as Primary and Secondary, such classifications being determined by the magnitude of current to be expected from electrostatically induced charges, and precipitation static charges respectively. Primary bonding conductors are used between major components, engines, external surfaces, e.g. flight control surfaces, and the main structure or earth. Secondary bonding conductors are used between components and earth for which primary conductors are not specifically required, e.g. pipelines carrying flammable fluids, metal conduits, junction boxes, door plates, etc.

Some static charge is always liable to remain on an aircraft so that after landing a difference in potential between the aircraft and the ground could be caused. This obviously is undesirable, since it creates an electric shock hazard to persons entering or leaving the aircraft, and can cause spark discharge between the aircraft and external ground equipment being coupled to it. In order to provide the necessary leakage path, two methods are generally adopted either separately or in combination. In one, the aircraft is fitted with a nosewheel or tail-wheel tyre as appropriate, the rubber of which contains a compound providing the tyre with good electrical conductivity. The second method provides a leakage path via short flexible steel wires secured to the nose wheel or main wheel axle members and making physical contact with the ground.

During refuelling of an aircraft, stringent precautions are necessary to minimize the risk of fire or explosion due to the presence of static charges. The aircraft itself may be charged, the fuel flowing through the hose generates electrical potentials, and the fuel tanker may be charged. Thus potential differences must be prevented from occurring and which could otherwise result in the generation of sparks and ignition of flammable vapours. The equalizing of potentials is achieved by providing a bonding connection between the aircraft and tanker which themselves are bonded to the ground, and by bonding the hose nozzle to a point specially provided on the aircraft. During the refuelling operation physical contact between the hose nozzle and tank filler is always maintained.

In a number of current "new technology" aircraft, high-performance non-metallic composite materials are used in major structural areas. Although they obviously have advantages from the structural point of view, the reduction in what may be termed the "metallic content" of the airframe does, unfortunately. reduce the effectiveness of shielding the aircraft, and electrical and electronic systems from the effects of lightning strikes.

Since the operation of these systems is based on sophisticated digital computing techniques, then the many computers in control of an aircraft are most vulnerable. This is because the magnetic induction created by large lightning currents flowing nearby, could completely destroy microprocessors and other vital integrated circuit packs comprising the units. Thus, further lightning protection must be built into the aircraft, and currently this takes the form of special suppression filters and metallic shielding over system cables.

STATIC DISCHARGE WICKS

As noted earlier, the discharge of static takes place continuously in order to equalize the potentials of the charges in the atmosphere and the aircraft. However, it is often the case that the rate of discharge is lower than the actual charging rate, with the result that the aircraft's charge potential reaches such a value it permits what is termed a corona discharge, a discharge which if of sufficient magnitude, will glow in poor visibility or at night. Corona discharge occurs more readily at curves and sections of an aircraft having minimum radii such as wing tips, trailing edges, propeller tips, horizontal and vertical stabilizers, radio antennae, pitot tubes, etc.

Corona discharge can cause serious interference with radio frequency signals and means must therefore be provided to ensure that the discharges occur at points where interference will be minimized. This is accomplished by devices called static discharge wicks or more simply, static dischargers. They provide a relatively easy exit for the charge so that the corona breaks out at predetermined points rather than haphazardly at points favourable to its occurrence. Static dischargers are fitted to the trailing edges of ailerons, elevators and rudder of an aircraft. A typical static discharger consists of nichrome wires formed in the manner of a "brush" or wick thereby providing a number of discharge points. In some instances, static dischargers may also take the form of small metal rods for trailing edge fitting and short flat metal blades for fitting at the tips of wings, horizontal and vertical stabilizers. Sharp tungsten needles extend at right angles to the discharger tips to keep corona voltage low and to ensure that discharge will occur only at these points.

SCREENING
Screening performs a similar function to bonding in that it provides a low resistance path for voltages producing unwanted radio frequency interference. However, whereas a bonding system is a conducting link for voltages produced by the build up of static charges, the voltages to be conducted by a screening system are those stray ones due to the coupling of external fields originating from certain items of electrical equipment, and circuits when in operation. Typical examples are: d.c. generators, engine ignition systems, d.c. motors, time switches and similar apparatus designed for making and breaking circuits at a controlled rate.

The methods adopted for screening are generally of three main types governed principally by the equipment or circuit radiating the interference fields. In equipment such as generators, motors and time switches several capacitors, which provide a low resistance path, are interconnected across the interference source, i.e. brushes, commutators and contacts, to form a self-contained unit known as a suppressor. The other methods adopted are the enclosing of equipment and circuits in metal cases and the enclosure of cables in a metal braided sheath, a method used for screening the cables of ignition systems. The suppressors and metal screens are connected to the main earth or ground system of an aircraft.

STANDARDIZING OF DISTRIBUTION
As mentioned in the introduction to this chapter, an organized form of power distribution throughout an aircraft is essential, and as an example of this we may conclude by considering the sequence illustrated in Figs. 5.17 to 5.21. The diagrams, although based on the B737 power distribution and control system, are generally representative of the standard approach adopted in other types of public transport aircraft.

Figure 5.17 shows the routing of the feeder lines

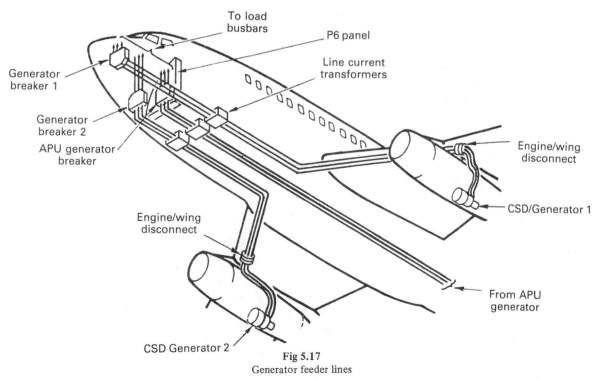

Fig 5.17
Generator feeder lines

96

from the main generators and the APU generator. At the wing/fuselage junction, the lines pass through sealed connectors into the underfloor area. All lines are then routed through an electrical/electronics compartment (see Fig. 5.18). Those from the main generators pass through sealed connectors into the unpressurized nosewheel well to connect up with the generator breakers. The feeder lines from the APU generator are connected to its breaker located above floor level within a special compartment (designated P6 panel) to the rear of the captain's position. This compartment contains most of the a.c. and d.c. busbars, the bus-tie breakers, voltage control and protection units for all three generators, and an external power control unit. The feeder lines from the main generator breakers pass into this compartment to connect with the a.c. busbars. A circuit breaker panel is mounted on the front side

of the compartment. As a complete unit therefore, the compartment or P6 panel, establishes what is termed the load control centre of the aircraft.

The electrical/electronics compartment serves as a centralized area for the rack mounting of the many "black boxes" associated with automatic flight control, compass, radio, and certain other airframe systems. Removal, installation and maintenance checks of these boxes are thereby facilitated by this arrangement.

In order to establish an organized form of systems control by each member of the flight crew, and also of circuit protection, an appropriate number of control panels are strategically located on the flight deck as illustrated in Figs. 5.19, 5.20, and 5.21. The panels are designated by the letter "P" prefixing the panel numbers, and in the example considered they are as follows:

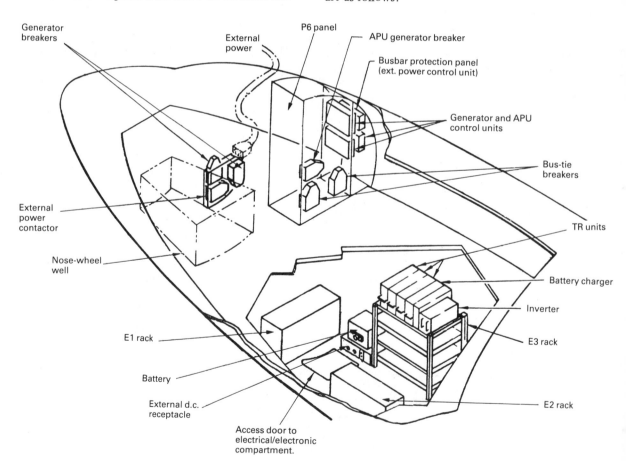

Fig 5.18 Electrical/electronics compartment

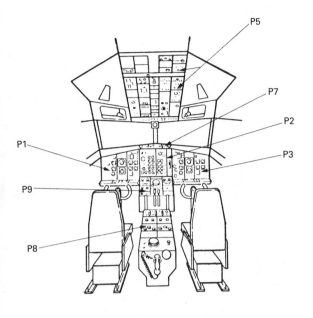

Fig 5.19
Flight deck control panels

LE DEVICES P5-12	THRUST REVERSER P5-30			OXYGEN P5-14	FLT REC P5-19
					STALL WARN P5-18
		OVHD		
FLIGHT CONTROLS FLT CONT P5-3	AC & DC METERING P5-13			ANTI-ICE WINDOW HEAT P5-9	TEMP CONTROL
	P5-5			ANTI-ICE P5-11	P5-17
P5-24	CSD & STBY ELEC			HYDRAULICS HYD P5-8	PNEUMATICS AIR COND
				DOORS DOOR WARN P5-20	
	BUS SWITCHING & GEN AMMS FUEL P5-4			P5-7	P5-10
P5-2	APU			PRESSURIZATION P5-16	P5-6
		P5-1		

Fig 5.20
Overhead control panel

P7   Glare shield panel containing annunciator lights for each pilot

P8   Fire protection system control panel

P9   Weather radar display indicator and radio communication system selector controls

    Circuit breakers are located on panels behind each pilot as shown in Fig. 5.21, and since they are allied to load control, then they are part of the load control centre, as for example, the P6 panel. The breakers are grouped appropriate to each system as indicated in Fig. 5.21.

P1   Captain's flight instrument panel

P2   Centre engine instrument panel

P3   First Officer's flight instrument panel

P5   Overhead panel from which electrical power generation systems and other major systems of the aircraft are controlled. Reference to Fig. 5.20 shows that the panel is subdivided into sections so that control switches, meters and indicating lights are grouped appropriate to their respective systems. For example, section P5–4 is the primary one for switching the main generators, APU generator and external power onto the busbars.

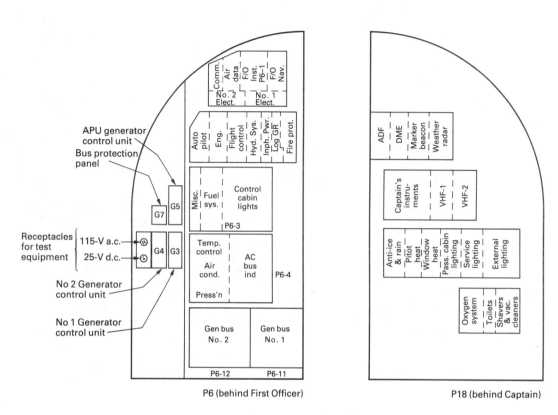

Fig 5.21
Circuit breaker panels (load control centres)

CHAPTER SIX

Circuit Controlling Devices

In aircraft electrical installations the function of initiating, and subsequently controlling the operating sequences of constituent circuits is performed principally by switches and relays, and the construction and operation of some typical devices form the subject of this chapter. It may be noted that although circuit breakers may also come within the above functional classification, they are essentially circuit protection devices and, as such, are separately described in the appropriate chapter.

Switches and relays are constructed in a variety of forms, and although not exhaustive, the details given in Table 6.1 may be considered a fairly representative summary of the types and the actuating methods commonly employed.

Switches

In its simplest form, a switch consists of two contacting surfaces which can be isolated from each other

Table 6.1

Switching Device	Primary method of actuating contact assemblies				Remarks
	Manual	Mechanical	Electrical	Electromagnetic	
SWITCHES					
Toggle	X				
Push	X			X	Certain types incorporate a "hold-in" coil; lights.
Rotary	X				
Micro	X	X	X		
Rheostat	X				
Time		X			Mechanical timing device operated in turn by an electric motor.
Mercury		X			
Pressure		X			
Thermal		X	X		Effects of metal expansion and also of electric current.
Proximity		X	X		
Solid-state			X		Transistor type "on-off". Used in the internal circuits of units such as control and protection.
RELAYS				X	Electromagnetic, in turn controlled by a circuit incorporating one or more manual switches, mechanical switches or a combination of these.
BREAKERS or Contactors				X	

or brought together as required by a movable connecting link. This connecting link is referred to as a *pole* and when it provides a single path for a flow of current as shown in Fig. 6.1(a), the switch is designated as a *single-pole, single-throw* switch. The term *throw* thus indicates the number of circuits each pole can complete through the switch. In many circuits, various switching combinations are usually required, and in order to facilitate the make and break operations, the contact assemblies of switches (and certain relays) may be constructed as integrated units. For example, the switch at (b) of Fig. 6.1 can control two circuits in one single make or break operation, and is therefore known as a *double-pole, single-throw* switch, the poles being suitably insulated from each other. Two further examples are illustrated in diagrams (c) and (d) and are designated *single-pole, double-throw* and *double-pole, double-throw* respectively.

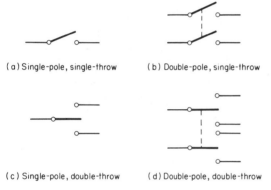

(a) Single-pole, single-throw

(b) Double-pole, single-throw

(c) Single-pole, double-throw

(d) Double-pole, double-throw

Fig 6.1
Switch contact arrangements

In addition to the number of poles and throws, switches (toggle types in particular) are also designated by the number of positions they have. Thus, a toggle switch which is spring-loaded to one position and must be held at the second to complete a circuit, is called a *single-position* switch. If the switch can be set at either of two positions, e.g. opening the circuit in one position and completing it in another, it is then called a *two-position* switch. A switch which can be set at any one of three positions, e.g. a centre "off" and two "on" positions, is a *three-position* switch, also known as a *selector* switch.

TOGGLE SWITCHES
Toggle or tumbler-type switches, as they are sometimes called, perform what may be regarded as "general-

purpose" switching functions and are used extensively in the various circuits. A typical switch is illustrated in Fig. 6.2.

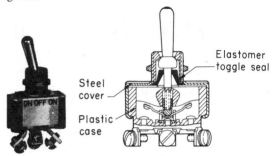

Fig 6.2
Toggle switch

In some applications it may be necessary for the switches in several independent circuits to be actuated simultaneously. This is accomplished by "ganging" the switches together by means of a bar linking each toggle as shown in Fig. 6.3(a). A variation of this method is used in certain types of aircraft for simultaneous action of switch toggles in one direction only

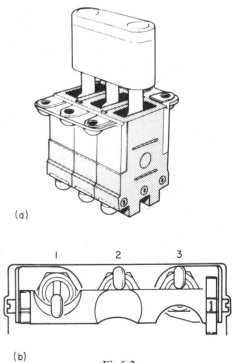

(a)

(b)

Fig 6.3
"Ganging" and locking of switches

(usually to a "system off" position). This is accomplished by a separate gang-bar mounted on the control panel in such a way that it can be pulled down to bear against the toggles of the switches to push them in the required direction. When the bar is released it is returned under the action of a spring.

A further variation is one in which the operation of a particular switch, or all in a series, may be constrained. A typical application to a triple generator system is shown in Fig. 6.3(b), the switches being used for the alternative disposition of busbar loads in the event of failure of any of the three generators.

A locking bar is free to rotate in mounting brackets anchored by the locking nuts of the No. 1 and No. 2 switches. The radiused cut-outs, at 90 degrees to each other, are provided along the length of the bar at positions coincident with the toggles of each switch. A steel spring provides for tensioning of the bar at each selected position, and is inserted around the circumference at the right-hand end. Markings 1, 2, 3 and "N" correspond to the positions of the cut-outs on the bar relative to the switch toggles. If, for example, there is a failure of No. 1 generator the bar is rotated to the position 1 permitting operation of failure switch No. 1, but constraining the toggles of the other two switches. The action for switch operation at positions 2 and 3 is similar. Thus, the busbar loads of a failed generator can be distributed between remaining serviceable generators at the same time avoiding inadvertent switch operation. When the letter "N" is evident the bar and the cut-outs are positioned so that none of the switches can be operated.

PUSH-SWITCHES

Push-switches are used primarily for operations of short duration, i.e. when a circuit is to be completed or interrupted momentarily, or when an alternative path is to be made available for brief periods. Other variants are designed to close one or more circuits (through separate contacts) while opening another circuit, and in these types, provision may be made for contact-action in the individual circuits to occur in sequence instead of simultaneously. In basic form a push-switch consists of a button-operated spring-loaded plunger carrying one or more contact plates which serve to establish electrical connection between fixed contact surfaces. Switches may be designed as independent units for either "push-to-make" or "push-to-break" operation, or designed to be double-acting. For certain warning and indicating purposes, some types contain miniature lamps positioned behind a

small translucent screen in the push-button. When illuminated, legends such as "on", "closed" or "fail" are displayed on the screen and in the appropriate colours.

The construction of a simple type of "push-to-make" switch and the arrangement of an illuminated type are shown in Fig. 6.4. In some circuits, for example in a turbopropeller engine starting circuit (see also p. 156), switches are designed to be both manual and electromagnetic in operation. A typical example, normally referred to as a "push-in solenoid

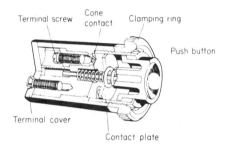

Simple type

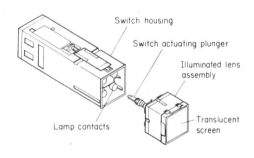

Illuminated type

Fig 6.4
Push switches

switch", is shown in Fig. 6.5. The components are contained within a casing comprising an aluminium housing having an integral mounting flange, a sleeve and an end cover. The solenoid coil is located at the flange-end of the housing, and has a plunger passing through it. One end of the plunger extends beyond the housing flange and has a knob secured to it, while the other end terminates in a spring-loaded contact assembly. A combined terminal and fixed contact block is attached to the end of the housing and is held in place by a knurled end cover nut.

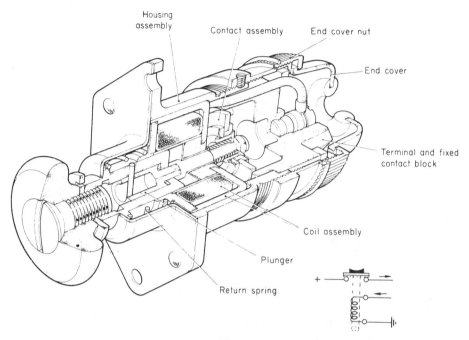

Fig 6.5
Push-in solenoid switch

When the plunger is depressed and held, the spring-loaded contact assembly bears against the fixed contacts and connects a d.c. supply to the starter motor. The commencement of the starting cycle provides a current flow through the hold-in coil of the switch, thereby energizing it and obviating the necessity for further manual control. The switch remains in the "on" position until the starting cycle is completed. At this stage, the current through the solenoid coil will have dropped sufficiently to permit the spring to return the plunger and contacts to the "off" position.

ROCKER-BUTTON SWITCHES
Rocker-button switches combine the action of both toggle and push-button type switches and are utilized for circuit control of some systems and equipment. A typical switch is shown in section in Fig. 6.6. For certain warning and indicating purposes, some types are provided with a coloured cap or screen displaying legend information, illuminated by a miniature lamp.

ROTARY SWITCHES
These are manually operated, and for certain operating requirements they offer an advantage over toggle switches in that they are less prone to accidental

operation. Furthermore, the rotary principle and positive engagement of contacts made possible by the constructional features make these switches more adaptable to multi-circuit selection than toggle type switches. A typical application is the selection of a single voltmeter to read the voltages at several bus-bars. In the basic form a rotary switch consists of a

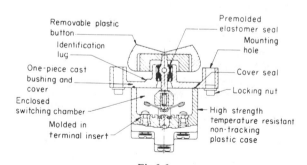

Fig 6.6
Rocker-button switch

central spindle carrying one or more contact plates or blades which engage with corresponding fixed contacts mounted on the switch base. The movement is usually spring-loaded and equipped with some form of eccentric device to give a snap action and positive engagement of the contact surfaces.

MICRO-SWITCHES

Micro-switches are a special category of switch and are one of the most extensively applied electrical devices in aircraft, performing a wide range of operations to ensure safe control of a variety of systems and components. The term "micro-switch" designates a switching device in which the differential travel between "make" and "break" of the operating mechanism is of the order of a few thousandths of an inch. Magnification and snap action of contact mechanism movements are derived from a pre-tensioned mechanically biased spring. The principle is shown in Fig. 6.7.

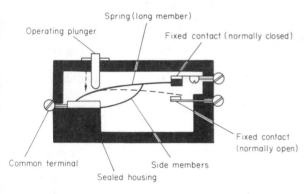

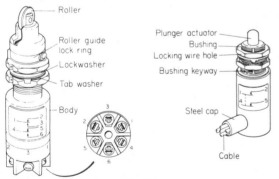

Fig 6.7
Micro-switches

The long member of the one-piece spring is cantilever supported and the operating button or plunger bears against the spring. Two shorter side members are anchored in such a way that they are bowed in compression. In the inoperative position the contact mounted on the free end of the spring is held against the upper fixed contact by the couple resulting from both tension and compression force. Depression of the operating button deflects the long member downwards thereby causing a reversal of the couple which "snaps" the spring and contact downward. Upon removal of the operating force, cantilever action restores the spring and contact system to its initial position with a snap action.

The method of actuating micro-switches depends largely on the system to which it is applied but usually it is either by means of a lever, roller or cam, these in turn being operated either manually or electrically. The operating cycle of a micro-switch is defined in terms of movement of the operating plunger. This has a specified amount of pre-travel, or free movement before the switch snaps over. Following the operating point, there is some over-travel, while on the return stroke some differential travel beyond the operating point is provided before the release action of the switch takes place. The contacts of the switches shown in Fig. 6.7 operate within sealed evacuated chambers filled with an inert gas, e.g. nitrogen.

RHEOSTATS

These are controlling devices containing a resistance the magnitude of which can be varied, thereby adjusting the current in the circuit in which it is connected. A typical example of this method of control is the one adopted for varying the intensity of instrument panel and certain cockpit lighting.

Rheostats normally adjust circuit resistance without opening the circuit, although in some cases, they are constructed to serve as a combined on-off switch and variable resistor.

TIME SWITCHES

Certain consumer services are required to operate on a pre-determined controlled time sequence basis and as this involves the switching on and off of various components or sections of circuit, switches automatically operated by timing mechanisms are necessary. The principle of time switch operation varies, but in general it is based on the one in which a contact assembly is actuated by a cam driven at constant speed by either a speed-controlled electric motor or a spring-driven escapement mechanism. In some specialized consumer services, switches which operate on a thermal principle are used. In these the contact assembly is operated by the distortion of a thermal element when the latter has been carrying a designed current for a pre-determined period.

An example of a motor-driven time switch unit is

shown in Fig. 6.8. It is designed to actuate relays which, in turn, control the supply of alternating current to the heating elements of a power unit de-icing system (see p. 170). Signals to the relays are given in repeated time cycles which can be of short or long duration corresponding respectively to "fast" and "slow" selections made on the appropriate system control switch.

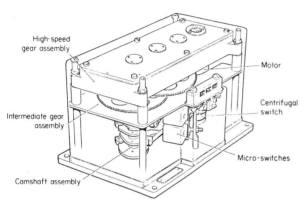

Fig 6.8
Time switch unit

The unit comprises an assembly of five cam and lever-actuated micro-switches driven by an a.c. motor through a reduction gearbox.

The motor runs at constant speed and drives the camshaft at one revolution per 240 seconds. Two of the cams are of the three-lobed type and they switch on two micro-switches three times during one revolution, each "on" period corresponding to 20 seconds. Two other cams are of the single-lobed type and they switch on two associated micro-switches once during one revolution, the "on" periods in this case corresponding to 60 seconds. Thus the foregoing cam and micro-switch operations correspond respectively to "fast" and "slow" selections of power to the heating elements, which are accordingly heated for short or long periods. The fifth cam and its micro-switch constitute what is termed a "homing" control circuit, the purpose of which is to re-set the time switch after use so that it will always re-commence at the beginning of an operating cycle.

When the "homing" micro-switch closes, it completes an external relay circuit whose function is to continue operation of the motor whenever the de-icing system is switched off. On completion of the full revolution of the camshaft, the homing micro-

switch is opened, thereby stopping the motor and resetting the timer for the next cycle of operation.

MERCURY SWITCHES

Mercury switches are glass tubes into which stationary contacts, or electrodes, and a pool of loose mercury are hermetically sealed. Tilting the tube causes the mercury to flow in a direction to close or open a gap between the electrodes to "make" or "break" the circuit in which the switch is connected.

The rapidity of "make" and "break" depends on the surface tension of the mercury rather than on externally applied forces. Thus, mercury switches are applied to systems in which the angular position of a component must be controlled within a narrow band of operation, and in which the mechanical force required to tilt a switch is very low. A typical application is in torque motor circuits of gyro horizons in which the gyros must be precessed to, and maintained in, the vertical position.

Mercury switches are essentially single-pole, single-throw devices but, as will be noted from Fig. 6.9, some variations in switching arrangements can be utilized.

PRESSURE SWITCHES

In many of the aircraft systems in which pressure measurement is involved, it is necessary that a warning be given of either low or high pressures which might constitute hazardous operating conditions. In some systems also, the frequency of operation may be such that the use of a pressure-measuring instrument is not justified since it is only necessary for some indication that an operating pressure has been attained for the period during which the system is in operation. To meet this requirement, pressure switches are installed in the relevant systems and are connected to warning or indicator lights located on the cockpit panels.

A typical switch is illustrated in Fig. 6.10. It consists of a metal diaphragm bolted between the flanges of the two sections of the switch body. As may be seen, a chamber is formed on one side of the diaphragm and is open to the pressure source. On the other side of the diaphragm a push rod, working through a sealed guide, bears against contacts fitted in a terminal block connected to the warning or indicator light assembly. The contacts may be arranged to "make" on either decreasing or increasing pressure, and their gap settings may be preadjusted in accordance with the pressures at which warning or indication is required.

Pressure switches may also be applied to systems

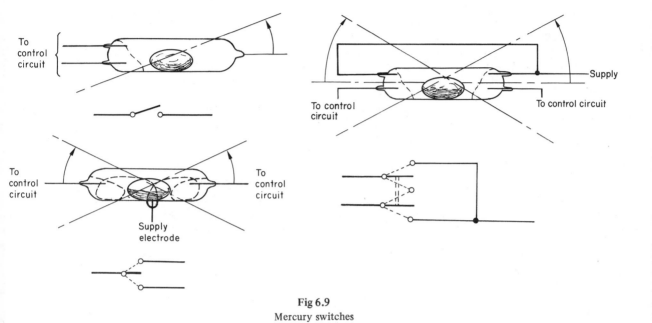

Fig 6.9
Mercury switches

Fig 6.10
Typical pressure switch unit

requiring that warning or indication be given of changes in pressure with respect to a certain datum pressure; in other words, as a differential pressure warning device. The construction and operation are basically the same as the standard type, with the exception that the diaphragm is subjected to a pressure on each side.

THERMAL SWITCHES

Thermal switches are applied to systems in which a visual warning of excessive temperature conditions, automatic temperature control and automatic operation of protection devices are required. Examples of such applications are, respectively, overheating of a generator, control of valves in a thermal de-icing system and the automatic operation of fire extinguishers.

A principle commonly adopted for thermal switch operation is based on the effects of differences of expansion between two metals, usually invar and steel. In some cases mercury contact switches may be employed.

An example of a differential expansion switch employed in some cases as a fire detecting device, is shown in Fig. 6.11. The heat-sensitive element is an alloy steel barrel containing a spring bow assembly of low coefficient of expansion. Each limb of the bow carries a silver-rhodium contact connected by fire-resistant cable to a terminal block located within a steel case.

In the event of a fire or sufficient rise in temperature at the switch location (a typical temperature is 300°C) the barrel will expand and remove the compressive force from the bow assembly, permitting the contacts to close the circuit to its relevant warning lamp. When the temperature drops, the barrel contracts, thus compressing the bow assembly and re-opening the contacts.

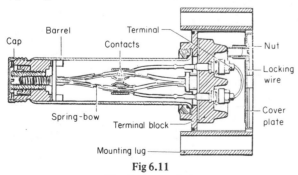

Fig 6.11
Fire detector switch

PROXIMITY SWITCHES

These switches are used in several types of aircraft as part of circuits required to give warning of whether or not passenger entrance doors, freight doors, etc. are fully closed and locked. Since they have no moving parts they offer certain advantages over micro-switches which are also applied to such warning circuits.

A typical switch shown in Fig. 6.12 consists of two

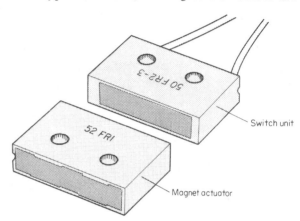

Fig 6.12
Proximity switch

main components, one of which is an hermetically-sealed permanent magnet actuator, and the other a switch unit comprising two reeds, each having rhodium-plated contacts connected to the warning circuit. The two components are mounted in such a manner that when they contact each other, the field from the permanent magnet closes the reeds and contacts together, to complete a circuit to the "door closed" indicator.

Relays

Relays are in effect, electromagnetic switching devices by means of which one electrical circuit can be indirectly controlled by a change in the same or another electrical circuit.

Various types of relay are in use, their construction, operation, power ratings, etc., being governed by their applications, which are also varied and numerous. In the basic form, however, a relay may be considered as being made up of two principal elements, one for sensing the electrical changes and for operating the relay mechanism, and the other for controlling the changes. The sensing and operating element is a solenoid and armature, and the controlling element is one or more pairs or contacts.

As in the case of switches, relays are also designated by their "pole" and "throw" arrangements and these can range from the simple single-pole, single-throw type to complex multiple contact assemblies controlling a variety of circuits and operated by the one solenoid.

In many applications the solenoid is energized directly from the aircraft power supply, while in others it may be energized by signals from an automatic device such as an amplifier in a cabin temperature-control system, or a fire detector unit. When the solenoid coil is energized a magnetic field is set up and at a pre-determined voltage level (called the "pull-in" voltage) the armature is attracted to a pole piece against spring restraint, and actuates the contact assembly, this in turn either completing or interrupting the circuit being controlled. When the solenoid coil circuit is interrupted at what is termed the "dropout" voltage, the spring returns the armature and contact assembly to the inoperative condition.

In addition to the contact assembly designations mentioned earlier, relays are also classified by the order of making and breaking of contacts, whether normally open ("NO") or normally closed ("NC") in the de-energized position, rating of the contacts in amperes and the voltage of the energizing supply. The design of a relay is dictated by the function it is required to perform in a particular system or component, and as a result many types are available, making it difficult to group them neatly into specific classes. On a very broad basis, however, grouping is usually related to the basic form of construction, e.g. attracted core, attracted-armature, polarized armature, and "slugged", and the current-carrying ratings of the controlling element contacts, i.e. whether heavy-duty or light-duty. The descriptions given in the following paragraphs are therefore set out on this basis and the relays selected are typical and generally representative of applications to aircraft systems.

ATTRACTED-CORE, HEAVY-DUTY RELAY
The designation "heavy-duty" refers specifically to the amount of current to be carried by the contacts. These relays are therefore applied to circuits involving the use of heavy-duty motors which may take starting currents over a range from 100 A to 1500 A, either short-term, as for starter motors for example, or continuous operation.

A relay of the type used for the control of a

typical turbopropeller engine starter motor circuit is illustrated in Fig. 6.13. The contact assembly consists of a thick contact plate and two suitably insulated fixed contact studs connected to the main terminals.

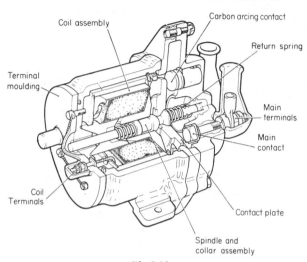

Fig 6.13
Attracted core heavy-duty relay

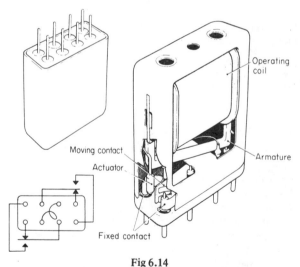

Fig 6.14
Attracted armature light-duty relay (sealed)

The contact plate is mounted on a supporting spindle and this also carries a soft inner core located inside the solenoid coil. The complete moving component is spring-loaded to hold the contact plate from the fixed contacts and to retain the core at the upper end of the coil. When the coil is energized the polarities of the magnetic fields established in the coil and core are such that the core moves downwards against spring pressure, until movement is stopped by the contact plate bridging across the fixed contacts, thus completing the main circuit. Carbon contacts are provided to absorb the initial heavy current and thereby reduce arcing to a minimum before positive connection with the main contacts is made.

ATTRACTED-ARMATURE, LIGHT-DUTY RELAY
A relay designed for use in a 28-volt d.c. circuit and having a contact rating of 3 A is shown in Fig. 6.14.

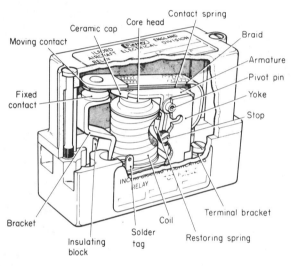

Fig 6.15
Attracted armature relay (unsealed type)

The contacts are of a silver alloy and are actuated in the manner shown in the inset diagram, by a pivoted armature. In accordance with the practice adopted for many currently used relays, the principal elements are enclosed in an hermetically-sealed case filled with dry nitrogen and the connection in the circuit is made via a plug-in type base. Fig. 6.15 illustrates another example of attracted armature relay. This is of the unsealed type and is connected into the relevant circuit by means of terminal screws in the base of the relay.

POLARIZED ARMATURE RELAYS
In certain specialized applications, the value of control circuit currents and voltages may be only a few milliamps and millivolts, and therefore relays of exceptional sensitivity are required. This requirement cannot always be met by relays which employ spring-controlled armatures, for although loading may be decreased to permit operation at a lower "pull-in" voltage, effective control of the contacts is decreased and there is a risk of contact flutter. A practical solution to this problem resulted in a relay in which the attraction and repulsion effects of magnetic forces are substituted for the conventional spring-control of the armature and contact assembly. Fig. 6.16 shows, in diagrammatic form, the essential features and operating principle of such a relay.

The armature is a permanent magnet and is pivoted between two sets of pole faces formed by a frame of high permeability material (usually mu-metal). It is lightly biased to one side to bring the contact assembly into the static condition as in Fig. 6.16(a). The centre limb of the frame carries a low-inductance low-current winding which exerts a small magnetizing force on the frame when it is energized from a suitable source of direct current. With the armature in the static condition, the frame pole-faces acquire, by induction from the armature, the polarities shown, and the resulting forces of magnetic attraction retain the armature firmly in position.

When a d.c. voltage is applied to the coil the frame becomes, in effect, the core of an electromagnet. The flux established in the core opposes and exceeds the flux due to the permanent magnet armature, and the frame pole-faces acquire the polarities shown in Fig. 6.16(b). As the armature poles and frame pole-faces are now of like polarity, the armature is driven to the position shown in Fig. 6.16(c) by the forces of repulsion. In this position it will be noted that poles and pole-faces are now of unlike polarity, and strong forces of attraction hold the armature and contact assembly in the operating condition. The fluxes derived from the coil and the armature act in the same direction to give a flux distribution as shown in Fig. 6.16(c). When the coil circuit supply is interrupted, the permanent magnet flux remains, but the force due to it is weaker than the armature bias force and so the armature and contacts are returned to the static condition (Fig. 6.16(a)).

SLUGGED RELAYS
For some applications requirements arise for the use of relays which are slow to operate the contact assembly either at the stage when the armature is being attracted, or when it is being released.

Some relays are therefore designed to meet these requirements, and they use a simple principle whereby the build-up or collapse of the main electromagnet

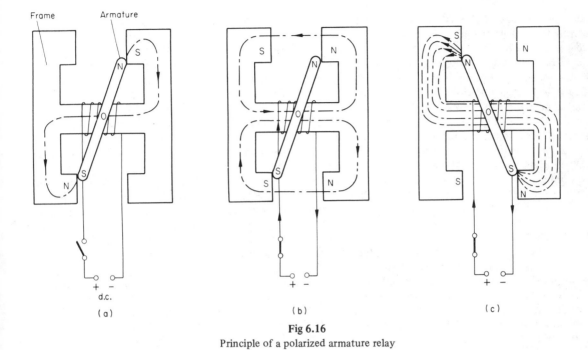

Fig 6.16

Principle of a polarized armature relay

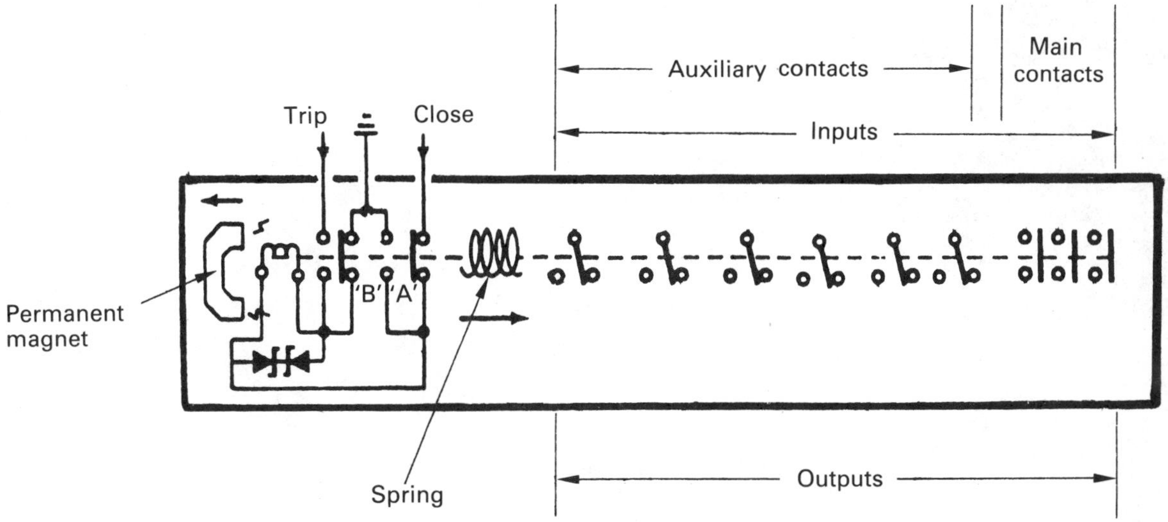

Frame   Armature

S
N
O
S
N

+ −
d.c.

(a)

S   N
O
S   N

+ −

(b)

S
N
O
S
N

+ −

(c)

Auxiliary contacts

Main contacts

Inputs

Trip   Close

Permanent magnet

'B' 'A'

Spring

Outputs

Fig 6.17
Breaker

flux is slowed down by a second and opposing magnetizing force. This procedure is known as "slugging" and a relay to which it is applied is called a "slug" relay. The relay usually incorporates a ring of copper or other non-magnetic conducting material (the "slug") in the magnetic circuit of the relay, in such a way that changes in the operating flux which is linked with the slug originate the required opposing magnetic force. In some slug relays the required result is obtained by fitting an additional winding over the relay core and making provision for short-circuiting the winding, as required, by means of independent contacts provided in the main contact assemblies.

BREAKERS

These devices sometimes referred to as contactors, are commonly used in power generation systems for the connection of feeder lines to busbars, and also for interconnecting or "tying" of busbars (see also page 78). The internal arrangement of one such breaker is shown in Fig. 6.17.

It consists of main heavy-duty contacts for connecting the a.c. feeder lines, and a number of smaller auxiliary contacts which carry d.c. for the control of other breakers, relays, indicating lights as appropriate to the overall system. All contacts are closed and/or tripped by a d.c.-operated electromagnetic coil; a permanent magnet serves to assist the coil in closing, and also to latch the breaker in the closed position. The coil is also assisted in tripping by means of a spring. Two zener diodes are connected across the coil to suppress arcing of the coil circuit contacts during closing and tripping.

When say, a main generator switch is placed in its "on" position, a d.c. "closing" signal will flow through the relaxed contacts "A" and then through the coil to ground via relaxed contacts "B". With the coil energized, the main and other auxiliary contacts will therefore be closed and the spring will be compressed. The changeover of the coil contacts "A" completes a hold-in circuit to ground, and with the assistance of the permanent magnet the breaker remains latched.

A tripping signal resulting from either the generator switch being placed to "off", or from a fault condition sensed by a protection unit, will flow to ground in the opposite direction to that when closing, and via the second set of the "close" contacts. The spring assists the reversed electromagnetic field of the coil in breaking the permanent magnetic latch.

Breakers of this type are installed with their opening-closing axis in the horizontal position.

CHAPTER SEVEN

Circuit Protection Devices and Systems

In the event of a short circuit, an overload or other fault condition occurring in the circuit formed by cables and components of an electrical system, it is possible for extensive damage and failure to result. For example, if the excessive current flow caused by a short circuit at some section of a cable is left unchecked, the heat generated in the cable will continue to increase until something gives way. A portion of the cable may melt, thereby opening the circuit so that the only damage done would be to the cable involved. The probability exists, however, that much greater damage would result; the heat could char and burn the cable insulation and that of other cables forming a loom, and so causing more short circuits and setting the stage for an electrical fire. It is essential therefore to provide devices in the network of power distribution to systems, and having the common purpose of protecting their circuits, cables and components. The devices normally employed are fuses, circuit breakers and current limiters. In addition, other devices are provided to serve as protection against such fault conditions as reverse current, overvoltage, undervoltage, overfrequency, underfrequency, phase unbalance, etc. These devices may generally be considered as part of main generating systems, and those associated with d.c. power generation, in particular, are normally integrated with the generator control units.

FUSES

A fuse is a thermal device designed primarily to protect the cables of a circuit against the flow of short-circuit and overload currents. In its basic form, a fuse consists of a low melting point fusible element or link, enclosed in a glass or ceramic casing which not only protects the element, but also localizes any flash which may occur when "fusing". The element is joined to end caps on the casing, the caps in turn, providing the connection of the element with the circuit it is designed to protect. Under short-circuit or overload current conditions, heating occurs, but before this can affect the circuit cables or other elements, the fusible element, which has a much lower current-carrying capacity, melts and interrupts the circuit. The materials most commonly used for the elements are tin, lead, alloy of tin and bismuth, silver or copper in either the pure or alloyed state.

The construction and current ratings of fuses vary, to permit a suitable choice for specific electrical installations and proper protection of individual circuits. Fuses are, in general, selected on the basis of the lowest rating consistent with reliable sytem operation, thermal characteristics of cables, and without resulting "nuisance tripping". For emergency circuits, i.e., circuits the failure of which may result in the inability of an aircraft to maintain controlled flight and effect a safe landing, fuses are of the highest rating possible consistent with cable protection. For these circuits it is also necessary that the cable and fuse combination supplying the power be carefully engineered taking into account short-term transients in order to ensure maximum utilization of the vital equipment without circuit interruption.

Being thermal devices, fuses are also influenced by ambient temperature variations. These can affect to some extent the minimum "blowing" current, as well as "blowing" time at higher currents, and so must also be taken in account. Typical examples of fuses currently in use in light and heavy-duty circuits, are shown in Fig. 7.1(a)-(b) respectively. The light-duty fuse is screwed into its holder (in some types a bayonet cap fitting is used) which is secured to the fuse panel by a fixing nut. The circuit cable is connected to terminals located in the holder, the terminals making contact with corresponding connections on the

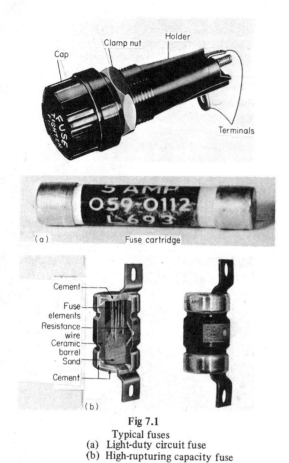

Fig 7.1
Typical fuses
(a) Light-duty circuit fuse
(b) High-rupturing capacity fuse

element cartridge. A small hole is drilled through the centre of the cap to permit the insertion of a fuse test probe.

Fuses are located accessible for replacement, and as close to a power distribution point as possible so as to achieve the minimum of unprotected cable.

The heavy-duty or high rupturing capacity fuse (Fig. 7.1(b)) is designed for installation at main power distribution points (by means of mounting lugs and bolts). It consists of a tubular ceramic cartridge within which a number of identical fuse elements in parallel are connected to end contacts. Fire-clay cement and metallic end caps effectively seal the ends of the cartridge, which is completely filled with a packing medium to damp down the explosive effect of the arc set up on rupture of the fusible elements. The material used for packing of the fuse illustrated is granular quartz; other materials suitable for this purpose are magnesite (magnesium oxide), kieselguhr, and calcium carbonate (chalk). When an overload current condition arises and each element is close to fusing point, the element to go first immediately transfers its load to the remaining elements and they, now being well overloaded, fail in quick succession.

In some transport aircraft, the fuseholders are of the self-indicating type incorporating a lamp and a resistor, connected in such a way that the lamp lights when the fusible element ruptures.

CURRENT LIMITERS
Current limiters, as the name suggests, are designed to limit the current to some pre-determined amperage value. They are also thermal devices, but unlike ordinary fuses they have a high melting point, so that their time/current characteristics permit them to carry a considerable overload current before rupturing. For this reason their application is confined to the protection of heavy-duty power distribution circuits.

A typical current limiter (manufactured under the name of "Airfuse") is illustrated in Fig. 7.2. It incorporates a fusible element which is, in effect, a single strip of tinned copper, drilled and shaped at each end to form lug type connections, with the central portion "waisted" to the required width to form the fusing area. The central portion is enclosed by a rectangular ceramic housing, one side of which is furnished with an inspection window which, depending on the type, may be of glass or mica.

Fig 7.2
Typical current limiter ("Airfuse")

LIMITING RESISTORS
These provide another form of protection particularly in d.c. circuits in which the initial current surge is very high, e.g. starter motor and inverter circuits, circuits containing highly-capacitive loads. When such

circuits are switched on they impose current surges of such a magnitude as to lower the voltage of the complete system for a time period, the length of which is a function of the time response of the generating and voltage regulating system. In order therefore to keep the current surges within limits, the starting sections of the appropriate circuits incorporate a resistance element which is automatically connected in series and then shorted out when the current has fallen to a safe value.

Figure 7.3 illustrates the application of a limiting resistor to a turbine engine starter motor circuit incorporating a time switch; the initial current flow may be as high as 1500 A. The resistor is shunted across the contacts of a shorting relay which is controlled by the time switch. When the starter push switch is operated, current from the busbar flows through the coil of the main starting relay, thus energizing it. Closing of the relay contacts completes a circuit to the time switch motor, and also to the starter motor via the limiting resistor which thus reduces the peak current and initial starting torque of the motor. After a pre-determined time interval, which allows for a build-up of engine motoring speed, the torque load on the starter motor decreases and the time switch operates a set of contacts which complete a circuit to the shorting relay. As will be clear from Fig. 7.3, with the relay energized the current from the busbar passes direct to the starter motor, and the limiting resistor is shorted out. When ignition takes place and the engine reaches what is termed "self-sustaining speed", the power supply to the starter motor circuit is then switched off.

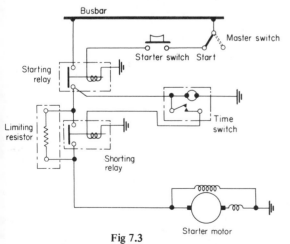

Fig 7.3
Application of a limiting resistor

CIRCUIT BREAKERS

Circuit breakers, unlike fuses or current limiters, isolate faulted circuits and equipment by means of a mechanical trip device actuated by the heating of a bi-metallic element through which the current passes to a switch unit. We may therefore consider them as being a combined fuse and switch device. They are used for the protection of cables and components and, since they can be reset after clearance of a fault, they avoid some of the replacement problems associated with fuses and current limiters. Furthermore, close tolerance trip time characteristics are possible, because the linkage between the bi-metal element and trip mechanism may be adjusted by the manufacturer to suit the current ratings of the element. The mechanism is of the "trip-free" type, i.e. it will not allow the contacts of the switch unit to be held closed while a fault current exists in the circuit.

The factors governing the selection of circuit breaker ratings and locations, are similar to those already described for fuses.

The design and construction of circuit breakers varies, but in general they consist of three main assemblies; a bi-metal thermal element, a contact type switch unit and a mechanical latching mechanism. A push-pull button is also provided for manual resetting after thermal tripping has occurred, and for manual tripping when it is required to switch off the supply to the circuit of a system. The construction and operation is illustrated schematically in Fig. 7.4. (At (a)

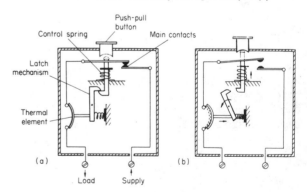

Fig 7.4
Schematic diagram of circuit breaker operation
(a) Closed
(b) Tripped condition

the circuit breaker is shown in its normal operating position; current passes through the switch unit contacts and the thermal element, which thus carries the

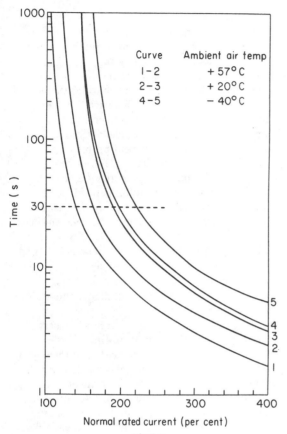

Fig 7.5

Characteristic curves of a typical circuit breaker tripping times

applied. The ambient temperature under which the circuit breaker operates also has an influence on circuit breaker operation and this, together with operating current values and tripping times, is derived from characteristic curves supplied by the manufacturer. A set of curves for a typical 6 A circuit breaker is shown in Fig. 7.5. The current values are expressed as a percentage of the continuous rating of the circuit breaker, and the curves are plotted to cover specified tolerance bands of current and time for three ambient temperatures. If, for example, the breaker was operating at an ambient temperature of +57°C, then in say 30 seconds it would trip when the load current reached a value between 140 and 160 per cent of the normal rating, i.e. between 8·4 and 9·6 A. At an ambient temperature of +20°C it would trip in 30 seconds at between 160 and 190 per cent of the normal rating (between 9·6 and 11·4 A) while at −40°C the load current would have to reach a value between 195 and 215 per cent of the normal rating (between 11·7 and 12·9 A) in order to trip in the same time interval.

After a circuit breaker has tripped, the distorted

full current supplied to the load being protected. At normal current values heat is produced in the thermal element, but is radiated away fairly quickly, and after an initial rise the temperature remains constant. If the current should exceed the normal operating value due to a short circuit, the temperature of the element begins to build up, and since metals comprising the thermal element have different coefficients of expansion, the element becomes distorted as indicated in Fig. 7.4(b). The distortion eventually becomes sufficient to release the latch mechanism and allows the control spring to open the switch unit contacts, thus isolating the load from the supply. At the same time, the push-pull button extends and in many types of circuit breaker a white band on the button is exposed to provide a visual indication of the tripped condition.

The temperature rise and degree of distortion produced in the thermal element are proportional to the value of the current and the time for which it is

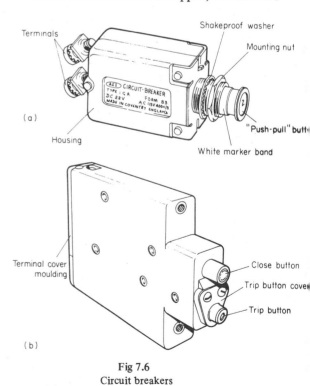

Fig 7.6
Circuit breakers
(a) Typical
(b) Circuit breaker with a "manual trip" button

element begins to cool down and reverts itself and the latch mechanism back to normal, and once the fault which caused tripping has been cleared, the circuit can again be completed by pushing in the circuit breaker button. This "resetting" action closes the main contacts and re-engages the push-button with the latch mechanism. If it is required to isolate the power supply to a circuit due to a suspected fault, or during testing, a circuit breaker may be used as a switch simply by pulling out the button. In some designs a separate button is provided for this purpose.

The external appearance of two typical single-pole, single-throw "trip-free" circuit breakers is illustrated in Fig. 7.6. The circuit breaker shown at (b) incorporates a separate manual trip push button. A cover may sometimes be fitted to prevent inadvertent operation of the button.

In three-phase a.c. circuits, triple-pole circuit breakers are used, and their mechanisms are so arranged that in the event of a fault current in any one or all three of the phases, all three poles will trip simultaneously. Similar tripping will take place should an unbalanced phase condition develop as a result of a phase becoming "open-circuited". The three trip mechanisms actuate a common push-pull button.

PROTECTION AGAINST REVERSE CURRENT
In all types of electrical systems the current flow is, of course, from the power source to the distribution busbar system and finally to the power consuming equipment; the interconnection throughout being made by such automatic devices as voltage regulators and control units, and by manually controlled switches. Under fault conditions, however, it is possible for the current flow to reverse direction, and as this would be of detriment to a circuit and associated equipment, it is therefore necessary to provide some automatic means of protection. In order to illustrate the fundamental principles we may consider two commonly used methods, namely reverse current relays and reverse current circuit breakers.

Reverse Current Cut-Out Relay

A reverse current cut-out relay is used principally in a d.c. generating system either as a separate unit or as part of a voltage regulator, e.g. the one described on p. 18. The circuit arrangement, as applied to the generating system typical of several types of small aircraft, is shown in Fig. 7.7. The relay consists of two coils wound on a core and a spring-controlled armature and contact assembly. The shunt winding is made up of many turns of fine wire connected across the generator so that voltage is impressed on it at all times. The series winding, of a few turns of heavy wire, is in series with the main supply line and is designed to carry the entire line current. The winding is also connected to the contact assembly, which under

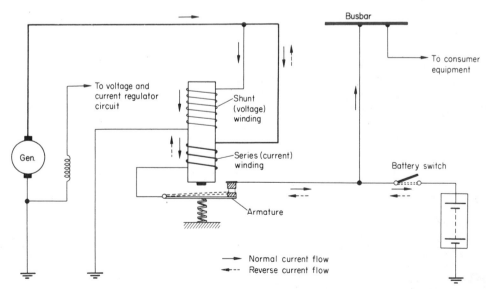

Fig 7.7
Reverse current cut-out operation

static conditions is held in the open position by means of a spring.

When the generator starts operating and the voltage builds up to a value which exceeds that of the battery, the shunt winding of the relay produces sufficient magnetism in the core to attract the armature and so close the contacts. Thus the relay acts as an automatic switch to connect the generator to the busbar, and also to the battery so that it is supplied with charging current. The field produced by the series winding aids the shunt-winding field in keeping the contacts firmly closed.

When the generator is being shut down or, say, a failure in its output occurs, then the output falls below the battery voltage and there is a momentary discharge of current from the battery; in other words, a condition of reverse current through the cut-out relay series winding is set up. As this also causes a reversal of its magnetic field, the shunt winding-field will be opposed, thereby reducing core magnetization until the armature spring opens the contacts. The generator is therefore switched to the "off-line" condition to protect it from damaging effects which would otherwise result from "motoring" current discharging from the battery.

Switched Reverse Current Relay

This relay is adopted in the d.c. generator systems of some types of small aircraft, its purpose being to permit switching of a generator on to the main bus-bar, and at the same time retain the disconnect function in the event of reverse current. The circuit arrangement is shown in Fig. 7.8.

In addition to a current coil the relay has a voltage coil, and a pair of contacts actuated via a contactor coil. When the voltage output is at a regulated value, the current through the voltage coil is sufficient to actuate its contacts which then connect the generator switch and contactor coil to ground. The contactor coil is thus energized from the A+ output of the generator and so the auxiliary and main contacts close to connect the generator output to the battery and main busbar. The magnetic effect of the current passing through the current coil assists that of the voltage coil in keeping the pilot contacts closed.

During engine shut-down, the generator output voltage decreases thereby initiating a reverse current condition, and because the magnetic effect of the current through the current coil now opposes that of the voltage coil, the pilot contacts open to de-

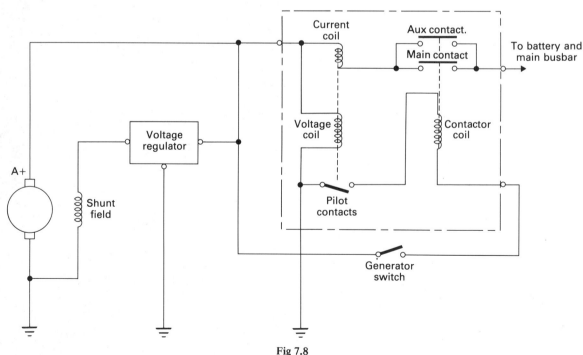

Fig 7.8
Switched reverse current relay

energize the contactor coil; thus, the main and auxiliary contacts are opened to disconnect the generator from the battery and main busbar.

Reverse Current Circuit Breakers

These circuit breakers are designed to protect power supply systems and associated circuits against fault currents of a magnitude greater than those at which cut-outs normally operate. Furthermore, they are designed to remain in a "locked-out" condition to ensure complete isolation of a circuit until a fault has been cleared.

An example of a circuit breaker designed for use in a d.c. generating system is shown in Fig. 7.9. It consists of a magnetic unit, the field strength and

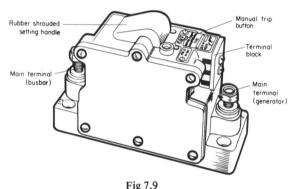

Fig 7.9
Reverse current circuit breaker

direction of which are controlled by a single-turn coil connected between the generator positive output and the busbar via a main contact assembly. An auxiliary contact assembly is also provided for connection in series with the shunt-field winding of the generator. The opening of both contact assemblies is controlled by a latching mechanism actuated by the magnet unit under heavy reverse current conditions. In common with other circuit breakers, resetting after a tripping operation has to be done manually, and is accomplished by a lever which is also actuated by the latching mechanism. Visual indication of a tripped condition is provided by a coloured indicator flag which appears behind a window in the circuit breaker cover. Manual tripping of the unit is effected by a push-button adjacent to the resetting lever.

Figure 7.10 is based on the circuit arrangement of a d.c. generating system used in a particular type of aircraft, and is an example of the application of a reverse

current circuit breaker in conjunction with a cut-out relay. Unlike the circuit shown in Fig. 7.7, the relay controls the operation of a line contactor connected in series with the coil of the reverse current circuit breaker. Under normal current flow conditions closing of the relay energizes the line contactor, the heavy-duty contacts of which connect the generator output to the busbar via the coil and main contacts of the normally closed reverse current circuit breaker. The magnetic field set up by the current flow assists that of the magnet unit, thus maintaining the breaker contacts in the closed position. The generator shunt field circuit is supplied via the auxiliary contacts.

When the generator is being shut down, or a failure of its output occurs, the reverse current resulting from the drop in output to a value below that of the battery flows through the circuit as indicated, and the cut-out relay is operated to de-energize the line contactor which takes the generator "off line". Under these conditions the reverse current circuit breaker will remain closed, since the current magnitude is much lower than that at which a specific type of breaker is normally rated (some typical ranges are 200–250 A and 850–950 A).

Let us consider now what would happen in the event of either the cut-out relay or the line contactor failing to open under the above low magnitude reverse current conditions, e.g. contacts have welded due to wear and excessive arcing. The reverse current would feed back to the generator, and in addition to its motoring effect on the generator, it would also reverse the generator field polarity. The reverse current passing through the circuit breaker coil would continue to increase in trying to overcome mechanical loads due to the engine and generator coupling, and so the increasing reverse field reduces the strength of the magnet unit. When the reverse current reaches the pre-set trip value of the circuit breaker, the field of the magnet unit is neutralized and repelled, causing the latch mechanism to release the main and auxiliary contacts to completely isolate the generator from the busbar. The breaker must be reset after the circuit fault has been cleared.

OVERVOLTAGE PROTECTION

Overvoltage is a condition which could arise in a generating system in the event of a fault in the field excitation circuit, e.g. internal grounding of the field windings or an open-circuit in the voltage regulator sensing lines. Devices are therefore necessary to pro-

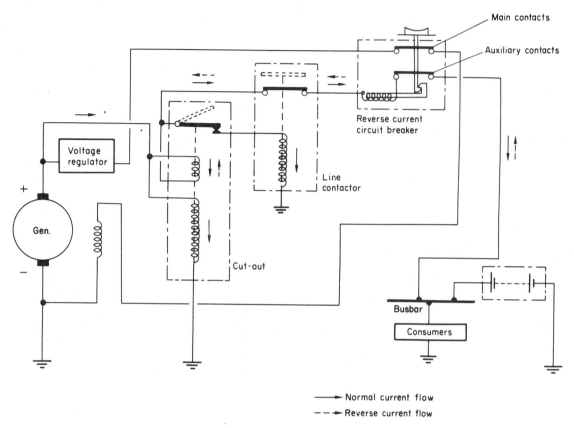

Fig 7.10
Reverse current circuit breaker operation

tect consumer equipment against voltages higher than those at which they are normally designed to operate. The methods adopted vary between aircraft systems and also on whether they supply d.c. or a.c. An example of an overvoltage relay method applied to one type of d.c. system is shown in Fig. 7.11.

The relay consists of a number of contacts connected in all essential circuits of the generator system, and mechanically coupled to a latching mechanism. This mechanism is electromagnetically controlled by a sensing coil and armature assembly, the coil being connected in the generator shunt-field circuit and in series with a resistor, the resistance of which decreases as the current through it is increased. Under normal regulated voltage conditions, the sensing coil circuit resistance is high enough to prevent generator shunt-field current from releasing the relay latch mechanism, and so the contacts remain closed and the generator remains connected to the busbar. If,

however, an open circuit occurs in the regulator voltage coil sensing line, shunt-field current increases and, because of the inverse characteristics of the relay sensing coil resistor, the electromagnetic field set up by the coil causes the latch mechanism to release all the relay contacts to the open position, thereby isolating the system from the busbar. After the fault has been cleared, the contacts are reset by depressing the push button.

Figure 7.12 illustrates a method employed in a frequency-wild a.c. generating system, the full control of which is provided by magnetic amplifiers (see also Chapter 2). The output of the overvoltage protection magnetic amplifier is fed to a bridge rectifier and to the coil of a relay, via a feedback winding. The main contacts of the relay are connected in the normal d.c. supply switching circuit to the line contactor.

Under normal voltage output conditions the impedance of the magnetic amplifier is such that its a.c.

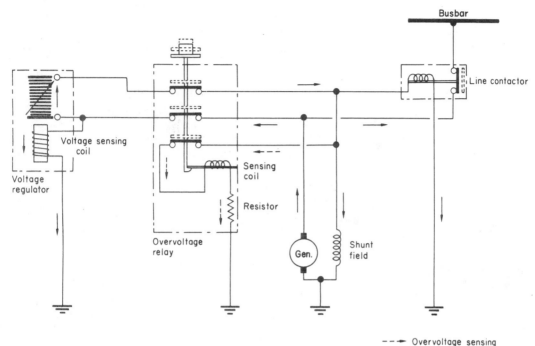

Fig 7.11
Overvoltage protection d.c. generating system

--- → Overvoltage sensing

output, and the rectified a.c. through the relay coil, maintain the relay in the de-energized condition. When an overvoltage condition is produced the current through the relay coil increases to a pre-determined energizing value, and the opening of the relay contacts interrupts the d.c. supply to the line contactor, which then disconnects the generator from the busbar. At the same time, the main control unit interrupts the supply of self-excitation current to the generator,

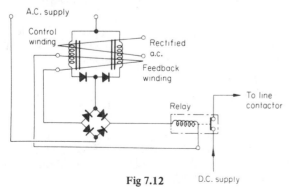

Fig 7.12
Overvoltage protection a.c. generating system
(frequency-wild)

causing its a.c. output to collapse to zero. The relay resets itself and after the fault has been cleared the generator output may be restored and connected to the busbar by carrying out the normal starting cycle.

An overvoltage protection system adopted in one example of a constant frequency (non-paralleled) a.c. generating system is shown in basic form in Fig. 7.13.

The detector utilizes solid-state circuit elements which sense all three phases of the generator output, and is set to operate at a level greater than 130 ± 3 volts. An overvoltage condition is an excitation-type fault probably resulting from loss of sensing to, or control of, the voltage regulator such that excessive field excitation of a generator is provided.

The signal resulting from an overvoltage is supplied through an inverse time delay to two solid-state switches. When switch S_1 is made it completes a circuit through the coil of the generator control relay, one contact of which opens to interrupt the generator excitation field circuit. The other contact closes and completes a circuit to the generator breaker trip relay, this in turn, de-energizing the

120

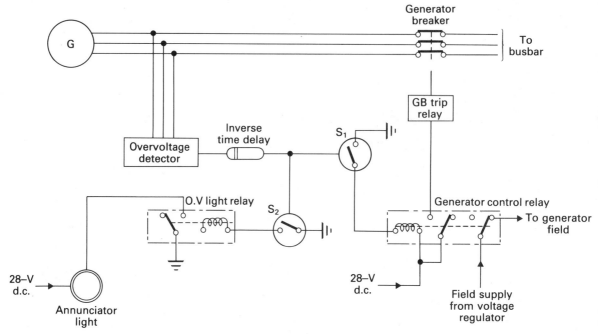

Fig 7.13
Overvoltage protection (constant frequency system)

generator breaker to disconnect the generator from the busbar. The making of solid-state switch S_2 energizes the light relay causing it to illuminate the annunciator light which is a white one in the actual system on which Fig. 7.13 is based. The purpose of the inverse time delay is to prevent nuisance tripping under transient conditions.

UNDERVOLTAGE PROTECTION

Undervoltage occurs in the course of operation when a generator is being shut down, and the flow of reverse current from the system to the generator is a normal indication of this condition. In a single d.c. generator system undervoltage protection is not essential since the reverse current is sensed and checked by the reverse current cut-out. It is, however, essential in a multi-generator system with an equalizing method of load-sharing, and since a load-sharing circuit always acts to raise the voltage of a lagging generator, then an undervoltage protection circuit is integrated with that of load-sharing. A typical circuit normally comprises a polarized relay which disconnects the load-sharing circuit and then allows the reverse current cut-out to disconnect the generator from the busbar.

In a constant frequency a.c. system, and considering the case of the one referred to on p. 78, the circuit arrangement for undervoltage protection is similar in many respects to that shown in Fig. 7.13, since it must also trip the generator control relay, the generator breaker, and must also annunciate the condition. The voltage level at which the circuit operates is less than 100 ± 3 volts. A time delay is also included and is set at 7 ± 2 seconds; its purpose being to prevent tripping due to transient voltages, and also to allow the CSD to slow down to an underfrequency condition on engine shutdown and so inhibit tripping of the generator control relay.

When generators are operating in parallel, undervoltage protection circuits are allied to reactive load-sharing circuits, an example of which was described on p. 50.

OVER-EXCITATION AND UNDER-EXCITATION PROTECTION

Over-excitation and under-excitation are conditions which are closely associated with those of overvoltage and undervoltage, and when generators are operating in parallel, the conditions are also associated with reactive current. Protection is therefore afforded by a mixing circuit. If the reactive current is the same in the generators paralleled, there will be no output from the circuit. When an unbalance occurs, e.g. a generator is over-excited, voltages will be produced in both over-excitation and under-excitation sections of the circuit, and these voltages will be fed to the overvoltage and undervoltage circuits. As a result, the overvoltage circuit will be biased down so that it will trip the generator breaker at a lower level. The undervoltage circuit will also be biased down so that it will trip the breaker at a lower voltage. Since the overall effect of over-excitation is to raise the busbar voltage then the overvoltage circuit provides the protective function.

With an under-excited generator, the voltages fed to the overvoltage and undervoltage circuits cause the biasing to have the opposite effect to over-excitation. Since under-excitation lowers the busbar voltage, then the undervoltage circuit provides the protective function.

UNDERFREQUENCY AND OVERFREQUENCY PROTECTION
Protection against these faults applies only to a.c. generating systems and is effected by the real load-sharing circuit of a generating system (see p. 48).

DIFFERENTIAL CURRENT PROTECTION
The purpose of a differential current protection system is to detect a short-circuited feeder line or generator busbar which would result in a very high current demand on a generator, and possibly result in an electrical fire. Under these conditions, the difference between the current leaving the generator and the current arriving at the busbar is called a differential fault or a feeder fault. In an a.c. system, current comparisons are made phase for phase, by two three-phase current transformers, one on the ground or neutral side of the generator (ground DPCT) and the other (the load DPCT) on the down-stream side of the busbar. Figure 7.14 illustrates the arrangement and principle of a system as applied to a single-phase line.

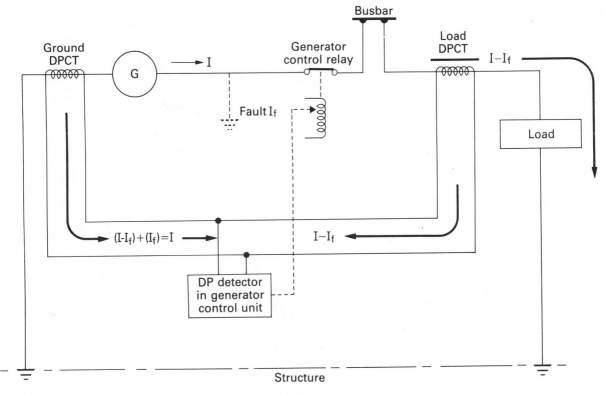

Fig 7.14
Differential current protection

If the current from the generator is I, and the fault current between the generator and busbar equals I_f, then the net current at the busbar will be equal to $I - I_f$. The fault current will flow through the aircraft structure and back to the generator through the ground DPCT. The remainder of the current $I - I_f$, will flow through the load DPCT, the loads, the aircraft structure, and then back to the generator via the ground DPCT. Thus, the ground DPCT will detect the generator's total current $(I - I_f) + (I_f)$ which is equal to I, and the load DPCT will detect $I - I_f$.

If the difference in current (i.e. the fault current) between the two current transformers on the phase line is sensed to be greater than the specified limit (20 or 30 amperes are typical values) a protector circuit within a generator control unit will trip the generator control relay.

MERZ–PRICE PROTECTION SYSTEM

This system is applied to some a.c. generating systems to provide protection against faults between phases or between one of the phases and ground. The connections for one phase are shown in Fig. 7.15; those for other phases (or other feeders in a single-phase system) being exactly the same. Two similar current transformers are connected to the line, one at each end, and their secondary windings are connected together via two relay coils. Since the windings are in opposition, and as long as the currents at each end of the line are equal, the induced e.m.f's are in balance and no current flows through the relay coils. When a fault occurs, the fault current creates an unbalanced condition causing current to flow through the coils of the relays thereby energizing them so as to open the line at each end.

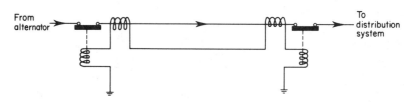

Fig 7.15
Merz–Price protection system

CHAPTER EIGHT

Measuring Instruments and Warning Indication Systems

In order to monitor the operating conditions of the various supply and utilization systems, it is necessary for measuring instruments and warning devices, in the form of indicators and lights, to be included in the systems. The number of indicating devices required and the types employed depend on the type of aircraft and the overall nature of its electrical installation. However, the layout shown in Fig. 8.1 is generally representative of systems monitoring requirements and can usefully serve as a basis for study of the appropriate indicating devices.

AMMETERS AND VOLTMETERS

These instruments are provided in d.c. and a.c. power generating systems and in most instances are of the permanent magnet moving-coil type shown in basic form in Fig. 8.2.

An instrument consists essentially of a permanent magnet with soft-iron pole pieces, between which a soft-iron core is mounted. A coil made up of a number of turns of fine copper wire is wound on an aluminium former which in turn is mounted on a spindle so that it can rotate in the air gap between the pole pieces and the core. The magnetic field in the air gap is an intense uniform radial field established by the cylindrical shape of the pole pieces and core. Current is led into and out of the coil through two hairsprings which also provide the controlling force. The hairsprings are so mounted that as the coil rotates, one spring is unwound and the other is wound. A pointer is attached to the spindle on which the moving coil is mounted.

When current flows through the coil a magnetic field is set up which interacts with the main field in the air gap such that it is strengthened and weakened as shown in the diagram. A force (Fd) is exerted on each side of the coil, and the couple so produced causes the coil to be rotated until it is balanced by

the opposing controlling force (Fc) of the hairsprings. Thus, rotation of the coil and pointer to the equilibrium position is proportional to the current flowing through the coil at that instant. This proportionality results in the evenly divided scale which is a characteristic of the moving coil type of indicator. When the coil former rotates in the main field, eddy currents are induced in the metal and these react with the main field producing a force opposing the rotation, thus bringing the coil to rest with a minimum of oscillation. Indicators of this kind are said to be "dead beat".

In order to protect the movements of these instruments against the effects of external magnetic fields and also to prevent "magnetic leakage", the movements are enclosed in a soft-iron case which acts as a magnetic screen. The soft-iron has a similar effect to the core of the indicator, i.e. it draws in lines of force and concentrates the field within itself.

Moving coil instruments are also generally employed for the measurement of voltage and current in an a.c. system. Additional components are necessary, of course, for each measuring application; e.g. for the measurement of voltage, the instrument must also contain a bridge rectifier while for the measurement of current, a shunt and a transformer are required in addition to the bridge rectifier.

Reference to Fig. 8.1 shows that all the instruments located on the control panel are of the circular-scale type; a presentation which is now adopted in many current types of aircraft. It has a number of advantages over the more conventional arc-type scale; namely, that the scale length is increased and for a given measuring range, the graduation of the scale can be more open, thus helping to improve the observational accuracy.

In order to cater for this type of presentation, it is, of course, necessary for some changes to be made in the arrangement of the magnet and moving coil

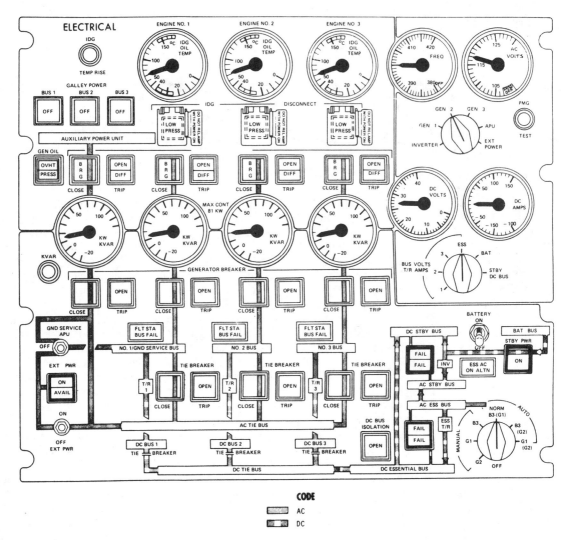

Fig 8.1
Electrical system control panel

systems, and one such arrangement is illustrated in Fig. 8.3.

The magnet is in the form of a block secured to a pole piece which is bored out to accommodate a core which itself is slotted and bored to permit the positioning of the moving coil. The coil former, unlike that of a conventional instrument, is mounted to one side of its supporting spindle, and under power-off conditions it surrounds the core and lies in the air gap at the position shown. The field flows from the magnet to the core which, in reality, forms a North pole, and then across the air gap to the pole piece forming the South pole. The return path of the field to the South pole of the magnet is completed through the yoke, which also shields the flux from distortion by external magnetic fields. When current flows through the coil, a force is produced due to the interaction between the permanent magnetic field and the induced field, but unlike the conventional instrument the coil is rotated about the core by a force acting on one side only; the opposite side being screened from the flux by the core itself.

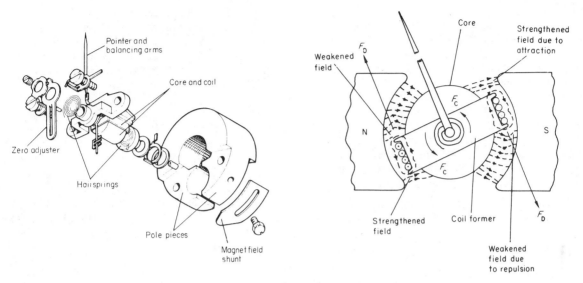

Fig 8.2
Basic form of moving coil indicator

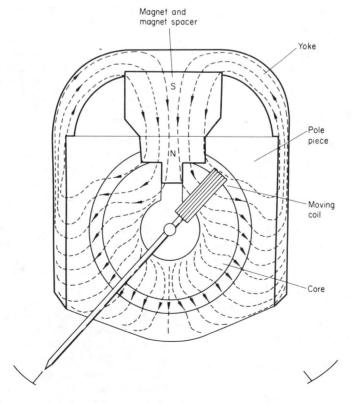

Fig 8.3
Magnet system of a typical long-scale moving coil instrument

SHUNTS

Shunts are used in conjunction with all d.c. system ammeters, and where specified, in a.c. systems, and their main purpose is to permit an ammeter to measure a large number of possible values of current, i.e. they act as range extension devices. Fundamentally, a shunt is a resistor having a very low value of resistance and connected external to the ammeter and in parallel with its moving coil. The materials used for shunts are copper, nichrome, manganin, minalpha and telcumen.

Typical shunts used in d.c. and a.c. generating systems are illustrated in Fig. 8.4 and although their principal physical features differ, a feature common to all shunts should be noted and that is they are

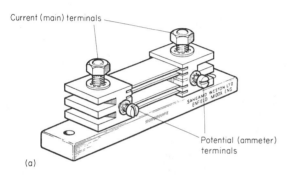

Current (main) terminals

Potential (ammeter) terminals

(a)

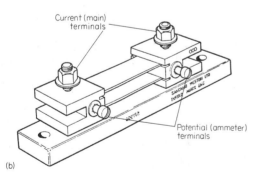

Current (main) terminals

Potential (ammeter) terminals

(b)

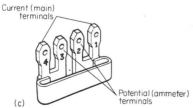

Current (main) terminals

Potential (ammeter) terminals

(c)

Fig 8.4
Shunts

each provided with four terminals. Two of these are of large current-carrying capacity ("current" terminals) for connecting the shunt in series with the main circuit, and two are of smaller size to carry smaller current ("potential" terminals) when connected to the associated ammeter. The unit shown at (a) employs strips of lacquered minalpha spaced from each other to promote a good circulation of air and thus ensure efficient cooling.

When the ammeter is in series with the main circuit only a fraction of the current passes through the moving coil, the remainder passing through the shunt which is selected to carry the appropriate load without overheating. The scale of the ammeter is, however, calibrated to indicate the range of current flow in the main circuit, since the flow through the coil and the shunt are in some pre-calculated ratio.

INSTRUMENT TRANSFORMERS

Transformers are used in conjunction with a.c. measuring instruments, and they perform a similar function to shunts, i.e. they permit a "scaling-down" of large currents and voltages to a level suitable for handling by standardized types of instruments. They fall into two main classes: (i) current or series transformers and (ii) potential or parallel transformers. The construction and operation of both classes has already been dealt with in Chapter 3 and at this stage therefore we shall only concern ourselves with typical applications.

Current transformers are normally used with a.c. ammeters and Fig. 8.5 illustrates a typical circuit arrangement. The main current-carrying conductor passes through the aperture of the secondary windings, the output of which is supplied to the ammeter via a bridge rectifier, which may be a separate unit or form part of the instrument itself.

An application of a potential transformer is illustrated in Fig. 8.6 and it will be noted that in this case the transformer forms part of a shunt, the primary winding being connected to the current terminals 1 and 4. The voltage developed across the shunt is stepped-up in the transformer to a maximum r.m.s. value (2·5 volts in this particular example) when the rated current is flowing through the shunt. The transformer output is connected to the "potential" terminals 2 and 3 and is rectified within the relevant ammeter and then applied to the moving coil. The scale of the ammeter used with this transformer arrangement is non-linear because the deflection of the moving coil is not proportional to the current

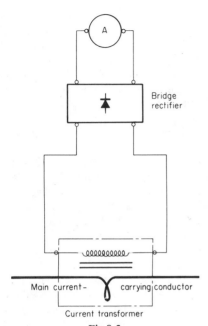

Fig 8.5

Application of a current transformer

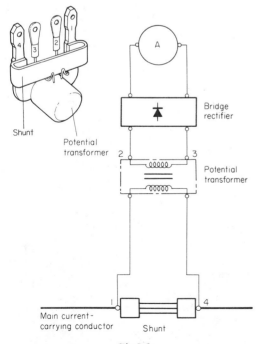

Fig 8.6

Application of a potential transformer

flowing through the shunt as a result of the sum of non-linear characteristics of the transformer and rectifier.

Figure 8.7 illustrates a circuit arrangement adopted for the measurement of d.c. loads in a rectified a.c. power supply system. The ammeter is utilized in conjunction with a three-phase current transformer, bridge rectifier and a shunt, which form

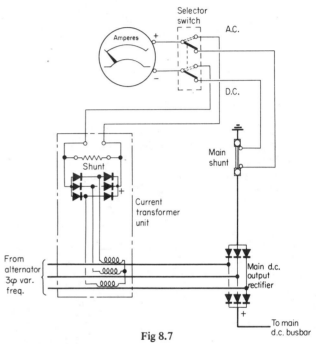

Fig 8.7

Measurement of d.c. loads in a rectified a.c. system

an integrated unit of the type shown in Fig. 8.8, and also a main shunt similar to that employed in basic d.c. generating systems. The ammeter is calibrated in amperes d.c. and it may be connected into either one of two circuits by means of a selector switch marked "D.C." and "A.C.". In the "D.C." position the ammeter is selected in parallel with the main shunt so that it measures the total rectified load taken from the main d.c. busbar.

When the "A.C." position is selected, the ammeter is connected to the shunt of the current transformer unit and as will be noted from the circuit diagram, this unit taps the generator output lines at a point before the main d.c. output rectifier. The transformer output is rectified for measuring purposes, so therefore in the "A.C." position of the switch, the ammeter will measure the d.c. equivalent of the total unrectified load.

128

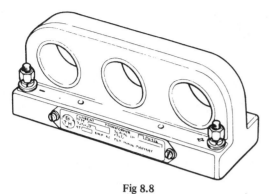

Fig 8.8
Three-phase current transformer unit

FREQUENCY METERS

These instruments form part of the metering system required for main a.c. power generating systems, and in some aircraft, they may also be employed in secondary a.c. generating systems utilizing inverters. The dial presentation and circuit diagram of a typical meter are shown in Fig. 8.9. The indicating

element, which is used in a mutual inductance circuit, is of the standard electrodynamometer pattern consisting essentially of a moving coil and a fixed field coil. The inductor circuit includes a nickel-iron core loading inductance, a dual fixed capacitor unit, four current-limiting resistors connected in series-parallel, and two other parallel-connected resistors which provide for temperature compensation. The electrical values of all the inductor circuit components are fixed.

The instrument also incorporates a circuit which is used for the initial calibration of the scale. The circuit is comprised of a resistor, used to govern the total length of the arc over which the pointer travels between the minimum and maximum frequencies, and a variable inductor system which governs the position of the centre of the arc of pointer travel relative to the mid-point of the instrument scale.

In operation the potential determined by the supply voltage and frequency is impressed on the field coil, which in turn sets up a main magnetic field in the area occupied by the moving coil. A second poten-

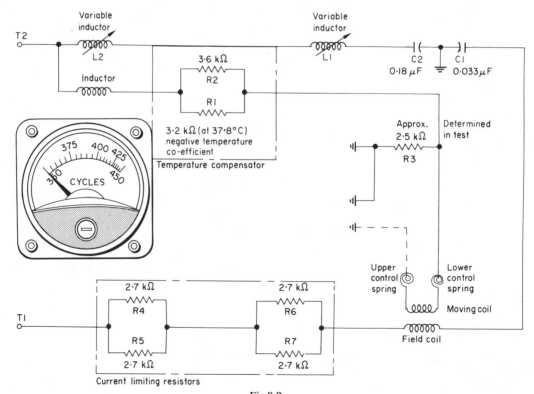

Fig 8.9
Circuit arrangements of a frequency meter

tial, whose value is also dependent on the supply voltage and frequency, is impressed on the moving coil, via the controlling springs. Thus, a second magnetic field is produced which interacts with the main magnetic field and also produces a torque causing the moving coil to rotate in the same manner as a conventional moving coil indicator. Rotation of the coil continues until the voltage produced in this winding by the main field is equal and opposite to the impressed potential at the given frequency. The total current in the moving coil and the resulting torque are therefore reduced to zero and the coil and pointer remain stationary at the point on the scale which corresponds to the frequency impressed on the two coils.

POWER METERS

In some a.c. power generating systems it is usual to provide an indication of the total power generated and/or the total reactive power. Separate instruments may be employed; one calibrated to read directly in watts and the other calibrated to read in var's (volt-amperes reactive) or, as in the case of the instrument illustrated in Fig. 8.10, both functions may be combined in what is termed a watt/var meter.

The construction and operation of the meter, not unlike the frequency meter described earlier, is based on the conventional electrodynamometer pattern and its scale, which is common to both units of measurement, is calibrated for use with a current transformer and an external resistor. A selector switch mounted adjacent to the meter provides for it to be operated as either a wattmeter or as a varmeter.

When selected to read in watts the field coil is supplied from the current transformer which as will be noted from Fig. 8.10 senses the load conditions at phase "B" of the supply. The magnetic field produced around the field coil is proportional to the load. The moving coil is supplied at 115 volts from phase B to ground and this field is constant under all conditions. The currents in both coils are in phase with each other and the torque resulting from both magnetic fields deflects the moving coil and pointer until balance between it and controlling spring torque is attained.

In the "var" position of the selector switch the field coil is again supplied from the current transformer sensing load conditions at phase "B". The moving coil, however, is now connected across phases "A" and "C" and in order to obtain the correct coil current, a calibrated resistor is connected in the circuit and

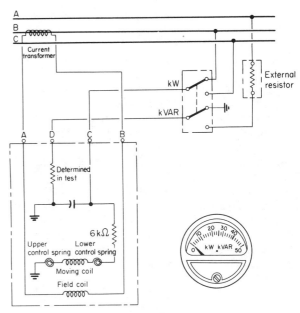

Fig 8.10
Circuit arrangements of a watt/VAR meter

mounted external to the instrument. The current in the moving coil is then at 90 degrees to the field coil current, and if the generator is loaded at unity power factor, then the magnetic fields of both coils bear the same angular relationship and no torque is produced.

For power factors less than unity there is interaction of the coil fields and a torque proportional to the load current and phase angle error is produced. Thus, the moving coil and pointer are rotated to a balanced position at which the reactive power is indicated.

WARNING AND INDICATING LIGHTS

Warning and indicator lights are used to alert the flight crew to conditions affecting the operation of aircraft systems. The lights may be divided into different categories according to the function they perform, and in general, we find that they fall into three main categories: (i) warning lights, (ii) caution lights and (iii) indicating or advisory lights.

Warning Lights. These are designed to alert the flight crew of unsafe conditions and are accordingly coloured red.

Caution Lights. These are amber in colour to indicate abnormal but not necessarily dangerous condi-

130

tions requiring caution, e.g. hydraulic system pressure running low.

Indicating or Advisory Lights. These lights, which are either green or blue, are provided to indicate that a system is operable or has assumed a safe condition, e.g. a landing gear down and locked.

Warning and indicator light assemblies are, basically, of simple construction, consisting of a bulb contained within a casing which incorporates electrical contacts and terminals for connection into the appropriate circuit. The coloured lens is contained within a cap which fits over the casing and bulb. Provision for testing the bulb to ensure that its filament is intact is also incorporated in many types of light assemblies. The lens cap is so mounted on the casing, that it can be pressed in to connect the bulb directly to the main power supply. Such an arrangement is referred to as a "press-to-test" facility.

Lights may also include a facility for dimming and usually this may be done in either of two ways. A dimming resistor may be included in the light circuit, or the lens cap may incorporate an iris type diaphragm which can be opened or closed by rotating the cap. Lights used for warning purposes do not usually include the dimming facility because of the danger involved in having a dimmed warning light escaping notice.

The power supplies for warning and indicator lights are derived from the d.c. distribution system and the choice of busbar for their connection must be properly selected. For example, if the failure of a system or a component is caused by the loss of supply to an auxiliary busbar, then it is obvious that if the warning light system is fed from the same busbar warning indications will also be lost. To avoid this risk it is necessary for warning lights to be supplied from busbars different from those feeding the associated service, and preferably on or as close as possible electrically to the busbar. Caution and indicating lights may also, in some cases, be supplied in a similar manner, but usually they are supplied from the same busbar as the associated service.

MAGNETIC INDICATORS

In many types of aircraft system, components require electrical control; for example, in a fuel system, electric actuators position valves which permit the supply of fuel from the main tanks to the engines and also for cross-feeding the fuel-supply. All such devices are, in the majority of cases, controlled by switches

on the appropriate systems panel, and to confirm the completion of movement of the device an indicating system is necessary.

The indicating system can either be in the form of a scale and pointer type of instrument, or an indicator light, but both methods can have certain disadvantages. The use of an instrument is rather space-consuming particularly where a number of actuating devices are involved, and unless it is essential for a pilot or systems engineer to know exactly the position of a device at any one time, instruments are uneconomical. Indicator lights are of course simpler, cheaper and consume less power, but the liability of their filaments to failure without warning contributes a hazard particularly in the case where "light out" is intended to indicate a "safe" condition of a system. Furthermore, in systems requiring a series of constant indications of prevailing conditions, constantly illuminated lamps can lead to confusion and misinterpretation on the part of the pilot or systems engineer.

Therefore to enhance the reliability of indication, indicators containing small electromagnets operating a shutter or similar moving element are installed on the systems panels of many present-day aircraft.

In its simplest form (see Fig. 8.11(a)) a magnetic indicator is of the two-position type comprising a ball pivoted on its axis and spring returned to the "off" position. A ferrous armature embedded in the ball is attracted by the electromagnet when energized, and rotates the ball through 150 degrees to present a different picture in the window. The picture can either be of the line diagram type, or of the instructive type.

Figure 8.11(b) shows a development of the basic indicator, it incorporates a second electromagnet which provides for three alternative indicating positions. The ferrous armature is pivoted centrally above the two magnets and can be attracted by either of them. Under the influence of magnetic attraction the armature tilts and its actuating arm will slide the rack horizontally to rotate the pinions fixed to the ends of prisms. The prisms will then be rotated through 120 degrees to present a new pattern in the window. When the rack moves from the centre "rest" position, one arm of the hairpin type centring spring, located in a slot in the rack, will be loaded. Thus, if the electromagnet is de-energized, the spring will return to mid-position rotating the pinions and prisms back to the "off" condition in the window.

The pictorial presentations offered by these indicators is further improved by the painting of "flow

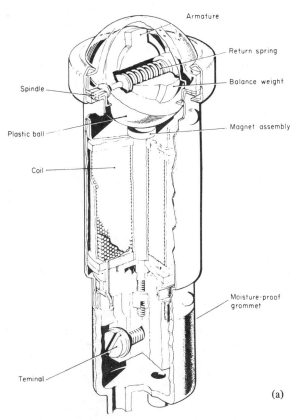

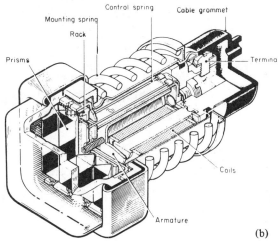

Fig 8.11
Magnetic indicators

lines" on the appropriate panels so that they interconnect the indicators with the system control switches, essential indicators and warning lights. A typical application of "flow lines" is shown in Fig. 8.1.

CENTRAL WARNING SYSTEMS

In the development of large types of aircraft and their associated systems, it became apparent that the use of warning and indicator lights in increasing numbers, and widely dispersed throughout flight compartments, would present a problem and that a new approach would be necessary. As a result, systems referred to as "central warning systems" were developed.

In its basic form, a system comprises a centralized group of warning and indicator lights connected to signal circuits actuated by the appropriate systems of the aircraft, each light displaying a legend denoting the system, and a malfunction or advisory message. All the lights are contained on an annunciator panel installed within a pilot's visual range.

An example of a system containing master warning and caution annunciator lights is shown in Fig. 8.12. The lights are centrally grouped according to systems, on a glare shield panel directly in front of the pilots and over their main instrument panels. The lights are also interconnected with systems indicating lights on an overhead control panel.

When a fault occurs in one of the systems, the overhead panel light for that system will illuminate, but as this may not always be readily observed by the pilots, their attention will be drawn to the fault situation by the simultaneous illumination of the annunciator light for the system, and of the master caution light. The lights are illuminated via a "fault pulser" and SCR circuit arrangement. Identification of the faulted system is cross-checked by observation of its control section of the overhead panel, and once this has been made, it is unnecessary for the master caution and annunciator lights to remain illuminated. They can therefore be extinguished by pressing the cap of either master caution light. If there is a need to recall the faulted system on an annunciator panel this can be accomplished by pressing the cap of the corresponding annunciator light. If the fault is not corrected a "recall pulser" circuit will retrigger the SCR and so illuminate the system annunciator light.

132

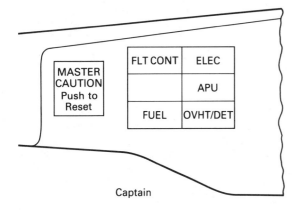

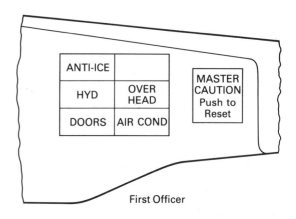

Fig 8.12
Master warning and caution lights

In aircraft carrying a flight engineer, a panel is also installed at his station and is functionally integrated with the pilot's panel. A flight engineer's panel is illustrated in Fig. 8.13 and may be taken as an example of central warning displays. In this case, the panel is made up of a number of blue lights which are advisory of normal operating conditions, a number of amber lights, a red "master warning" light and an amber "master caution" light.

When a fault occurs in a system, a fault-sensing device transmits a signal which illuminates the appropriate amber light. The signal is also transmitted to an electronic device known as a logic controller, the function of which is to determine whether the fault is of a hazardous nature or is one requiring caution. If the fault is hazardous, then the controller output signal illuminates the red "master warning" light; if caution is required, then the signal will illuminate only the amber "master caution" light.

Each master light incorporates a switch unit so that when the caps are pressed in, the active signal circuits are disconnected to extinguish the lights and, at the same time, they are reset to accept signals from faults which might subsequently occur in any other of the systems in the aircraft. The system lights are not of the resetting type and remain illuminated until the system fault is corrected. Dimming of lights and testing of bulb filaments is carried out by means of switches mounted adjacent to the annunciator panel.

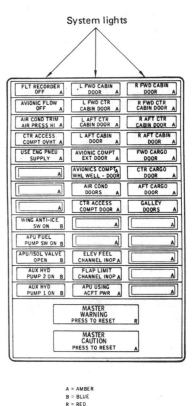

Fig 8.13
Centralized warning system annunciator panel

ELECTRONIC DISPLAY SYSTEMS

An electronic display system is one in which the data necessary for the in-flight operation of aircraft and systems, and also for their maintenance, is processed by high-storage capacity computers and then presented on the "screens" of colour cathode ray tube display units in alphanumeric and symbolic form. Advisory messages relating to faults in any one of the systems can also be displayed in this manner. Systems of this type are now in use in various types of aircraft, both military and civil, and in the latter category we may note as examples, the Airbus A310, Boeing 757 and 767.

The data for computer processing and for display, originates as signals (analogue and/or digital) generated by sensors associated with each individual major system of an aircraft, and as is conventional, data presentation falls into two broad areas: (i) flight path and navigational, and (ii) engine and airframe systems operation. Electronic display systems are therefore designed appropriate to each of these areas and are known respectively as an Electronic Flight Instrument System (EFIS) and either an Electronic Centralized Aircraft Monitor (ECAM) system or an Engine Indicating and Crew Alerting System (EICAS).

As far as electrical systems are concerned, the operational monitoring is handled by either an ECAM system or EICAS, and is confined to electrical power generation. The operation of both systems is understandably of a complex nature and space precludes detailed descriptions of them. We may however, gain some idea of their function, particularly in the role of fault annunciation by briefly referring to the ECAM system. For readers who may be interested in more details, reference may be made to "Microelectronics in Aircraft Systems" which is a companion volume to this.

A schematic functional diagram of the ECAM system (as used in the Airbus A310) is shown in Fig. 8.14. The CRT display units are mounted side by side so that the left-hand unit is dedicated to information in message form on systems' status, warnings and corrective action required, while the right-hand unit is dedicated to associated information in diagrammatic form. There are four modes of display, three of which are automatically selected and referred to as: flight phase-related, advisory and failure-related. The fourth mode is manual and permits the selection of diagrams related to any of the aircraft's systems, for routine checking and also the selection of status messages. The selections are made on the ECAM control panel.

In the context of this chapter, the failure-related mode is appropriate and an example of a display presentation is shown in Fig. 8.15. In this case, there is a problem associated with the number one generator. The left-hand display unit shows the affected system in message form, and in red or amber depending on the degree of urgency, and also the corrective action required in blue. At the same time, a diagram is displayed on the right-hand display unit. When the number one generator has been switched off, the light in the relevant push-button switch on the flight deck overhead panel is illuminated, and simultaneously, the blue instruction on the left-hand display unit changes to white. The diagram on the right-hand display unit is also "re-drawn" to depict by means of an amber line that the number one generator is no longer available, and that number two generator is supplying the busbar system. This is displayed in green which is the normal operating colour of the displays. After corrective action has been taken, the message on the left-hand display unit can be removed by operating a "clear" button switch on the ECAM control panel.

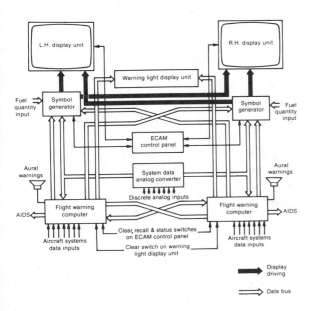

Fig 8.14
Schematic functional diagram − ECAM system

134

L.H. Display Unit
Display when warning detected

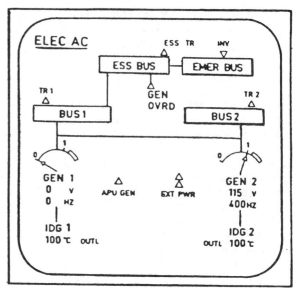

R.H. Display Unit

Fig 8.15
Display in failure-related mode

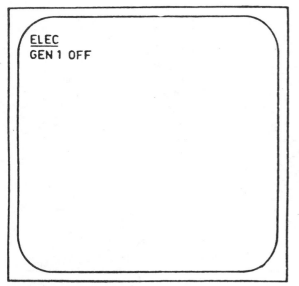

Display when corrective action taken

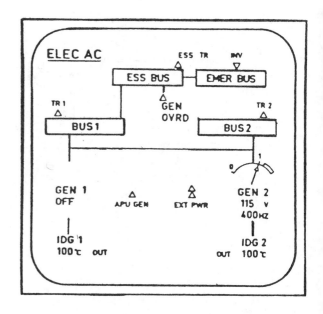

Fig 8.16
CRT display unit locations (Airbus A310)

Figure 8.16 shows the location of display units on the instrument panels of the Airbus A310. The units of the ECAM system are those occupying the central position, while EFIS units are on the left and right panels. There is also a third system which utilizes the electronic display concept, and that is known as a Flight Management System (FMS) the units of which are on the pedestal between the two pilots.

CHAPTER NINE

Power Utilization– Motors

Our study of electrical systems thus far, has been concerned primarily with the fundamental principles of the methods by which power is produced and distributed, and also of the circuit protection methods generally adopted. This study, however, cannot be concluded without learning something of the various ways in which the power is utilized within aircraft. Utilization can extend over very wide areas depending as it does on the size and type of aircraft, and whether systems are employed which require full or only partial use of electrical power; therefore, in keeping with the theme of the book, we shall only concern ourselves with some typical aspects and applications.

For the purpose of explanation, the subject is treated in this Chapter and in Chapter 10 respectively under two broad headings: (i) *motors* used in conjunction with mechanical systems, e.g. a motor-driven fuel valve; and (ii) *systems* which are principally all-electric, e.g. an engine starting and ignition system.

Motors

A wide variety of components and systems depend upon mechanical energy furnished by motors and the numbers installed in any one type of aircraft depend on the extent to which electrical power is in fact utilized. A summary of some typical applications of motors is given in Table 9.1.

In most of the above applications the motors and mechanical sections of the equipment form integrated units. The power supply required for operation is 28 volts d.c. and/or 26-volts or 115-volts constant frequency a.c. and is applied almost without exception, by direct switching and without any special starting equipment. Many motors are required to operate only for a short time during a flight, and ratings between 15 and 90 seconds are common. After operation at

the rated load, a cooling period of as long as 10 to 20 minutes may be necessary in some cases, e.g. a propeller feathering pump motor.

Table 9.1

	Function
Actuators	Fuel "trimming"; Cargo door operation; Heat exchanger control-flap operation; Landing flap operation.
Control Valves	Hot and cold air mixing for air-conditioning and thermal de-icing; Fuel shut-off.
Pumps	Fuel delivery; Propeller feathering; De-icing fluid delivery; Hydraulic fluid.
Flight Instruments and Control Systems	Gyroscope operation; Servo control

Continuously-rated motors are often fan cooled and, in the case of fuel booster pumps which are of the immersed type, heat is transferred from the sealed motor casing to the fuel. Operating speeds are high and in cases where the energy from motors must be converted into mechanical movements, reduction gear-boxes are used as the transmission system.

D.C. Motors

The function and operating principle of d.c. motors is the reverse of generators, i.e. if an external supply is connected to the terminals it will produce motion of the armature thereby converting electrical energy into mechanical energy. This may be seen from Fig. 9.1 which represents a motor in its simplest form, i.e. a single loop of wire "AB" arranged to rotate between the pole pieces of a magnet. The ends of the wire are connected to commutator segments which are contacted by brushes supplied with d.c. With

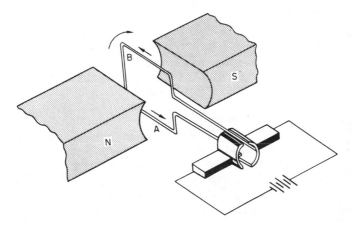

Fig 9.1
D.C. motor principle

current flowing in the loop in the direction shown, magnetic fields are produced around the wire which interact with the main field and produce forces causing the loop to move in a clockwise direction. When the loop reaches a position at which the commutator reverses the polarity of the supply to the loop, the direction of current flow is also reversed, but due to the relative positions of the field around the wire and of the main field at that instant, the forces produced cause the loop to continue moving in a clockwise direction. This action continues so long as the power is supplied to the loop.

As far as construction fundamentals are concerned, there is little difference between d.c. generators and motors; they both consist of the same essential parts, i.e. armature, field windings, commutator and brush-gear, the same methods of classifying according to various field excitation arrangements, and in the majority of motors the armature and field windings are supplied from a common power source, in other words they are self-excited.

MOTOR CHARACTERISTICS

The application of a motor to a particular function is governed by two main characteristics; the *speed characteristic* and the *torque characteristic*. The former refers to the variation of speed with armature current which is determined by the back e.m.f., this, in its turn, being governed by the mechanical load on the motor. The torque characteristic is the relationship between the torque required to drive a given load and the armature current.

TYPES OF MOTOR

There are three basic types of motors and as in the case of generators they are classified according to field excitation arrangements; series-wound, shunt-wound and compound-wound. These arrangements and certain other variations are adopted for a number of the functions listed in Table 9.1 and are illustrated in Fig. 9.2.

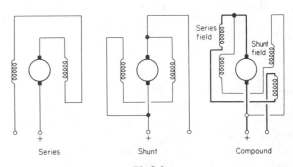

Fig 9.2
Types of d.c. motor

In *series-wound* motors, the field windings and the armature windings are connected in series with each other and the power supply. The currents flowing through both windings and the magnetic fields produced are therefore the same. The windings are of low resistance, and so a series motor is able to draw a large current when starting thereby eliminating building up the field strength quickly and giving the motor its principal advantages: high starting torque and good acceleration, with a rapid build-up of back e.m.f. induced in the armature to limit the current flow through the motor.

The speed characteristic of a series wound motor is such that variations in mechanical load are accompanied by substantial speed variations; a light load causing it to run at high speed and a heavy load causing it to run at low speed.

The torque is proportional to the square of the armature current, and as an increase in load results in a reduction of the back e.m.f., then there is an increase in armature current and a rapid increase in driving torque. Thus the torque characteristic is such that a motor can be started on full load.

In *shunt-wound* motors the field windings are arranged in the same manner as those of generators of this type, i.e. in parallel with the armature. The resistance of the winding is high and since it is connected directly across the power supply, the current through it is constant. The armature windings of some motors are of relatively high resistance and although their overall efficiency is low compared to the majority of shunt motors, they can be started by connecting them directly to the supply source. For the starting of motors having low-resistance armature windings it is necessary for a variable resistance to be connected in series with the armature. At the start full resistance would be in circuit to limit the armature current to some predetermined value. As the speed builds up the armature current is reduced by the increase in back e.m.f. and then the resistance is progressively reduced until, at normal speed of the motor, all resistance is out of the armature circuit.

In operating from a "no-load" to a "full-load" condition the variation in speed of a motor with a low-resistance armature is small and the motor can be considered as having a constant-speed characteristic. In the case of a motor with a high-resistance armature there is a more noticeable difference in speed when operating over the above load conditions.

The torque is proportional to the armature current until approaching full-load condition when the increase in armature reaction due to full-load current has a weakening effect. Starting torque is small since the field strength is slow to build up; thus, the torque characteristic is such that shunt-wound motors must be started on light or no load.

COMPOUND MOTORS

For many applications it is necessary to utilize the principal characteristics of both series and shunt motors but without the effects of some of their normally undesirable features of operation. For example, a motor may be required to develop the high starting torque of a series type but without the tendency to race when load is removed. Other applications may require a motor capable of reducing speed with increased load to an extent sufficient to prevent excessive power demand on the supply, while still retaining the smooth speed control and reliable "off-load" running characteristic of the shunt motor. These and other requirements can be met by what is termed *compounding,* or in other words, by combining both series and shunt field windings in the one machine. In most compound-wound motors the series and shunt windings are wound to give the same polarity on the pole faces so that the fields produced by each winding assist each other. This method of connection is known as cumulative compounding and there are three forms which may be used; normal, stabilized shunt and shunt limited.

In *normal compounding* a motor is biased towards the shunt-wound type, the shunt winding producing about 60 to 70 per cent of the total flux, the series winding producing the remainder. The desired characteristics of both series and shunt-wound motors are retained.

In the *stabilized shunt* form of compounding a motor is also biased towards the shunt-wound type but has a relatively minor series winding. The purpose of this winding is to overcome the tendency of a shunt motor to become unstable when running at or near its highest speed and then subjected to an increase in load.

The *shunt-limited* motor is biased towards the series-wound type and has a minor shunt field winding incorporated in the field system. The purpose of the winding is to limit the maximum speed when running under "off-load" conditions while leaving the torque and general speed characteristics unaltered. Shunt limiting is applied only to the larger sizes of compound motors, typical examples being engine starter motors (see Fig. 9.3). The speed/load characteristics of series, shunt and compound motors are shown in Fig. 9.4.

SPLIT-FIELD MOTORS

In a number of applications involving motors it is required that the direction of motor rotation be reversed in order to perform a particular function, e.g. the opening and closing of a valve by an actuator. This is done by reversing the direction of current flow

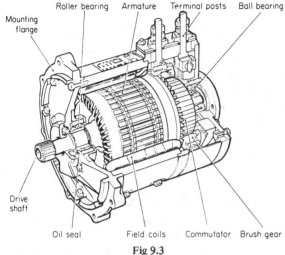

Fig 9.3
Typical starter motor

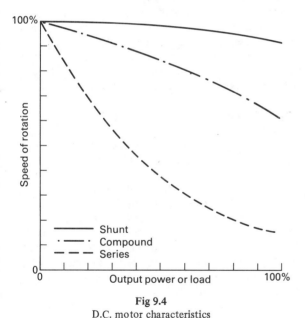

Fig 9.4
D.C. motor characteristics

circuit is shown in Fig. 9.5. When the switch is placed in the "Forward" position then current will flow in section "A" of the field winding and will establish a field in the iron core of appropriate polarity. Current also flows through the armature winding, the interaction of its field with that established by field winding section "A" causing the armature to

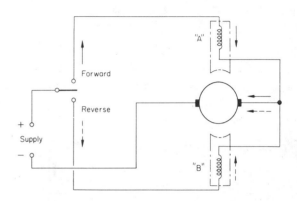

Fig 9.5
Split field motor circuit

rotate in the forward direction. When "Reverse" is selected on the control switch, section "A" is isolated and current flows through section "B" of the field winding in the opposite direction. The current flow through the armature is in the same direction as before, but as the polarities of the iron core pole pieces are now reversed then the resultant interaction of fields causes the armature to run in the reverse direction. Some split-field series motors are designed with two separate field windings on alternate poles. The armature in such a motor, a 4-pole reversible motor, rotates in one direction when current flows through the windings of one set of opposite pole pieces, and in the reverse direction when current flows through the other set of windings.

The reversing of motors by interchanging the armature connections is also employed in certain applications, notably when the operating characteristics of compound machines are required. The circuit diagram illustrated in Fig. 9.6 is based on the arrangement adopted in a compound motor designed for the lowering and raising of an aircraft's landing flaps (see Fig. 9.7). Current flows to the armature winding via the contacts of a relay, since the current demands of the motor are fairly high.

and magnetic field polarity, in either the field windings or the armature.

A method based on this principle, and one most commonly adopted in series-wound motors, is that in which the field winding is split into two electrically separate sections thereby establishing magnetic fields flowing in opposite directions. One of the two windings is used for each direction of rotation and is controlled by a single-pole double-throw switch. The

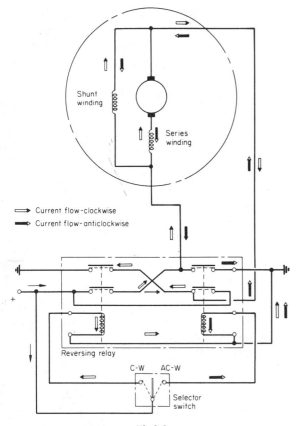

Fig 9.6
Reversing of a compound motor

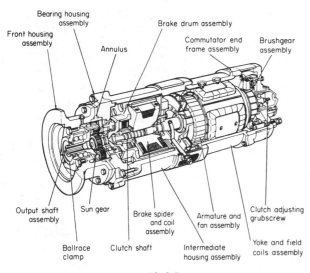

Fig 9.7
Reversible compound motor

Motor Actuators

Motor actuators are self-contained units combining electrical and mechanical devices capable of exerting reversible linear thrust over short distances, or reversible low-speed turning effort. Actuators are thereby classified as either linear or rotary and may be powered by either d.c. or a.c. motors. In the majority of cases d.c. motors are of the split-field series-wound type.

LINEAR ACTUATORS
Linear actuators may vary in certain of their design and constructional features dependent upon the application, load requirements and the manufacturer responsible. In general, however, they consist of the motor which is coupled through reduction gearing to a lead screw which on being rotated extends or retracts a ram or plunger. Depending on the size of actuator, extension and retraction is achieved either by the action of a conventional screw thread or by what may be termed a "ball bearing thread". In the former case, the lead screw is threaded along its length with a square-form thread which mates with a corresponding thread in the hollow ram. With the motor in operation the rotary motion of the lead screw is thereby converted into linear motion of the ram, which is linked to the appropriate movable component.

The ball bearing method provides a more efficient thread and is usually adopted in large actuators designed for operation against heavy loads. In this case, the conventional male and female threads are replaced by two semi-circular helical grooves, and the space between the grooves is filled with steel balls. As the lead screw rotates, the balls exert thrust on the ram, extending or retracting it as appropriate, and at the same time, a recirculating device ensures that the balls are fed continuously into the grooves.

A typical linear actuator is shown in Fig. 9.8.

ROTARY ACTUATORS
Rotary actuators are usually utilized in components the mechanical elements of which are required to be rotated at low speed or through limited angular travel. As in the case of linear actuators the drive from the motor is transmitted through reduction gearing, the output shaft of which is coupled directly to the relevant movable component, e.g. valve flap. Some typical examples of the application of rotary actuators are air-conditioning system spill valves and fuel cocks.

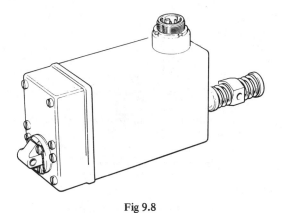

Fig 9.8
Linear actuator

LIMIT SWITCHES

Both linear and rotary type actuators are equipped with limit switches to stop their respective motors when the operating ram or output shaft, as appropriate, has reached the permissible limit of travel. The switches are of the micro type (see p. 104) and are usually operated by a cam driven by a shaft from the actuator gear-box. In some cases, limit switch contacts are also utilized to complete circuits to indicator lights or magnetic indicators. The interconnection of the switches is shown in Fig. 9.9, which is based on the circuit of a typical actuator-controlled valve system.

In the valve closed position, the cam operates the micro switch "A" so that it interrupts the "close" winding circuit of the motor and completes a circuit to the "closed" indicator. The contacts of the micro switch "B" are at that moment connected to the "open" winding of the motor so that when the control switch is selected, power is supplied to the winding. In running to the valve open position the cam causes micro switch "A" contacts to change over, thereby interrupting the indicator circuit and connecting the "close" winding so that the motor is always ready for operation in either direction. As soon as the "open"

ACTUATOR GEARING

The reduction gearing generally takes the form of multi-stage spur gear trains for small types of linear and rotary actuators, while in the larger types it is more usual for epicyclic gearing to be employed. The gear ratios vary between types of actuator and specific applications.

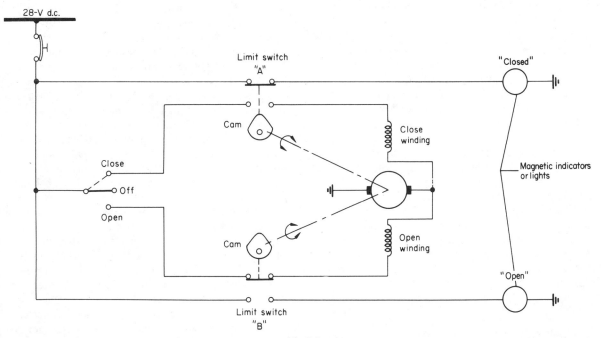

Fig 9.9
Limit switch operation

142

position is reached the cam operates micro switch "B", the contacts of which then complete a circuit to the "open" indicator.

BRAKES

The majority of actuators are fitted with electro-magnetic brakes to prevent over-travel when the motor is switched off. The design of brake system varies with the type and size of the actuator, but in all cases the brakes are spring-loaded to the "on" condition when the motor is de-energized, and the operating solenoids are connected in series with the armature so that the brakes are withdrawn immediately power is applied.

CLUTCHES

Friction clutches, which are usually of the single-plate type or multi-plate type dependent on size of actuator, are incorporated in the transmission systems of actuators to protect them against the effects of mechanical over-loading.

INSTRUMENT MOTORS

D.C. motors are not widely used in aircraft instruments, and in present-day systems they are usually confined to one or two types of turn-and-bank indicator to form the gyroscopic element. The motor armature together with a concentrically mounted outer rim forms the gyroscope rotor, the purpose of the rim being to increase the rotor mass and radius of gyration. The armature rotates inside a cylindrical two-pole permanent magnet stator secured to the gimbal ring. Current is fed to the brushes and commutator via flexible springs to permit gimbal ring movement. An essential requirement for operation of the instrument is that a constant rotor speed be maintained. This is achieved by a centrifugal cut-out type governor consisting of a fixed contact and a movable contact, normally held closed by an adjusting spring, and in series with the armature winding. A resistor is connected in parallel with the contacts.

When the maximum speed is attained, centrifugal force acting on the movable contact overcomes the spring restraint, causing the contacts to open. Current to the armature therefore passes through the resistor and so reduces rotor speed until it again reaches the nominal value.

A.C. Motors

In aircraft employing constant-frequency alternating current either as the primary or secondary source of electrical power, it is of course logical to utilize a.c. motors, and although they do not always serve as a complete substitute for d.c. machines, the advantages and special operating characteristics of certain types are applied to a number of systems which rely upon mechanical energy from an electromotive power source.

The a.c. motor most commonly used is the induction type, and dependent upon the application may be designed for operation from a three-phase, two-phase or single-phase power supply.

INDUCTION MOTORS

An induction motor derives its name from the fact that current produced in the rotating member, or rotor, is due to *induced e.m.f.* created by a rotating magnetic field set up by a.c. flowing in the windings of the stationary member or stator. Thus, interconnection between the two members is solely magnetic and as a result there is no necessity for a commutator, slip rings and brushes.

The rotor consists of a cylindrical laminated-iron core having a number of longitudinal bars of copper or aluminium evenly spaced around the circumference. These bars are joined at either end by copper or aluminium rings to form a composite structure commonly called a "squirrel-cage". The stator consists of a number of ring-shaped laminations having slots formed on the inner surface and into which series-connected coil windings are placed. The number of windings and their disposition within the stator is directly related to the number of poles and phases of the power supply, e.g. more windings are required in a 4-pole motor than in a 2-pole motor both of which are to be operated from a 3-phase supply.

The operating principle may be understood from Fig. 9.10, which represents a 2-pole 3-phase motor arrangement. Assuming that the relationship between phases (phase rotation) is as indicated, then at the instant 0°, phases "A" and "C" are the only two carrying current and they set up magnetic fields which combine to form a resultant field acting downward through the rotor core. The field thus passes the bars of the squirrel-cage, and since they form a closed circuit of low resistance the e.m.f. induced in the bars sets up a relatively large current flow in the direction indicated. As a result of the current flow magnetic fields are produced around the bars, each field interacting with the main field to produce torques on the rotor. This action is, in fact, the same as that which takes place in a d.c. motor and also a moving coil indicator.

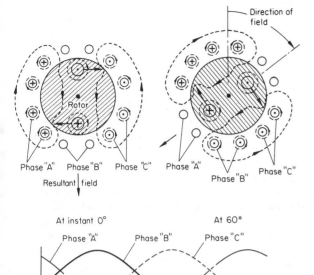

Fig 9.10
Induction motor principle

Assuming now that the power supply frequency has advanced through 60 degrees, then phase "A" current falls to zero, phases "B" and "C" are the two now carrying current and so the resultant field produced also advances through 60 degrees. In other words, the field starts rotating in synchronism with the frequency and establishes torques on the rotor squirrel-cage bars, thereby turning the rotor in the same direction as the rotating field in the stator. This action continues throughout the complete power supply cycle, the field making one complete revolution. In the case of a 4-pole motor the field rotates only 180 degrees during a full cycle and a 6-pole motor only 90 degrees.

As the speed of the rotor rises, there is a corresponding decrease of induced e.m.f. and torque until the latter balances the torques resulting from bearing friction, wind-resistance, etc., and the speed remains constant. Thus, the rotor never accelerates to the synchronous speed of the stator field; if it were to do so the bars would not be cut by the rotating field, there would be no induced e.m.f. or current flow, and no torque to maintain rotation.

The synchronous speed of an induction motor is determined by the number of poles for which the stator is wound, and the frequency of the power supply, i.e.

$$\text{Synchronous Speed} \atop \text{(rev/min)} = \frac{f\,(\text{Hz}) \times 60}{\text{No. of pairs of poles}}$$

The difference between the synchronous and rotor speeds, measured in rev/min., is called the *slip speed* and the ratio of this speed to synchronous speed, expressed as a percentage, is called quite simply the *slip*.

SINGLE-PHASE INDUCTION MOTORS

As the name indicates these motors have only one stator winding, and as this alone cannot produce a rotating field to turn the rotor then some other method of self-starting is necessary. The method most commonly adopted is the one in which the main winding of the stator is split to produce a second starting winding. Thus we obtain what is usually called a split-phase motor, and by displacing the windings mutually at 90 electrical degrees, and arranging that the current in the starting winding either leads or lags on the main winding, a rotating field can be produced in the manner of a two-phase motor. After a motor has attained a certain percentage of its rated speed, the starting winding may be switched out of the circuit; it then continues to run as a single-phase motor.

The lagging or leading of currents in the windings is obtained by arranging that the ratio of inductive reactance to resistance of one winding differs considerably from that of the other winding. The variations in ratio may be obtained by one of four methods, namely resistance starting, inductance starting, resistance/inductance starting or capacitance starting; the application of each method depends on the power output ratings of the particular motor. For example, horsepower ratings of capacitance starting motors are usually fractional and less than 2 h.p.

The first three methods are used only during starting of a motor, because if both windings remained in circuit under running conditions, the performance would be adversely affected. Moreover, a motor is able to run as a single-phase machine once a certain speed has been reached. The starting winding circuit is normally disconnected by a centrifugal switch. The fourth method can be used for both starting and running, and with suitably rated capacitors the running performance of capacitor motors, as they are called, approaches that of two-phase motors.

Figure 9.11 illustrates the application of a squirrel-cage capacitor motor to an axial-flow blower designed for radio rack cooling or general air circulation. It utilizes two capacitors connected in parallel and operates from a 115-volts single-phase 400 Hz supply. The capacitive reactance of the capacitors is greater than the inductive reactance of the starting winding, and so the current through this winding

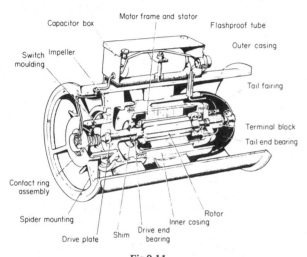

Fig 9.11
Motor-driven blower

thereby leads the supply voltage. The current in the running winding lags on the supply voltage and the phase difference causing field rotation is therefore the sum of the lag and lead angles.

TWO-PHASE INDUCTION MOTORS
These find their greatest applications in systems requiring a servo control of synchronous devices, e.g. as servomotors in power follow-up synchro systems.

The windings are also at 90 degrees to each other but, unlike the motors thus far described, they are connected to different voltage sources. One source is the main supply for the system and being of constant magnitude it serves as a reference voltage; the other source serves as a control voltage and is derived from a signal amplifier in such a way that it is variable in magnitude and its phase can either lead or lag the reference voltage, thereby controlling the speed and direction of rotation of the field and rotor.

HYSTERESIS MOTORS
Hysteresis motors also consist of a stator and rotor assembly, but unlike other a.c. motors the operation is directly dependent on the magnetism induced in the rotors and on the hysteresis or lagging characteristics of the material (usually cobalt steel) from which they are made.

A rotating field is produced by the stator and if the rotor is stationary, or turning at a speed less than the synchronous speed, every point on the rotor is subjected to successive magnetizing cycles. As the stator field reduces to zero during each cycle, a certain amount of flux remains in the rotor material, and since it lags on the stator field it produces a torque at the rotor shaft which remains constant as the rotor accelerates up to the synchronous speed of the stator field. This latter feature is one of the principal advantages of hysteresis motors and for this reason they are chosen for such applications as autopilot servomotors, which produce mechanical movements of an aircraft's flight control surfaces.

When the rotor reaches synchronous speed, it is no longer subjected to successive magnetizing cycles and in this condition it behaves as a permanent magnet.

CHAPTER TEN

Power Utilization – Systems

Lighting

Lighting plays an important role in the operation of an aircraft and many of its systems, and in the main falls into two groups: external lighting and internal lighting. Some of the principal applications of lights within these groups are as follows:

External Lighting

(i)   The marking of an aircraft's position by means of navigation lights.
(ii) Position marking by means of flashing lights.
(iii) Forward illumination for landing and taxi-ing.
(iv) Illumination of wings and engine air intakes to check for icing.
(v)   Illumination to permit evacuation of passengers after an emergency landing.

Internal Lighting

(vi)   Illumination of cockpit instruments and control panels.
(vii) Illumination of passenger cabins and passenger information signs.
(viii) Indication and warning of system operating conditions.

EXTERNAL LIGHTING

The plan view of external lighting given in Fig. 10.1 is based on the Boeing 747 and, although not all the lights shown would be standard on all other types of aircraft, it serves to illustrate the disposition of external lights generally.

NAVIGATION LIGHTS

The requirements and characteristics of navigation lights are agreed on an international basis and are set out in the statutory Rules of the Air and Orders for Air Navigation and Air Traffic Control regulations. Briefly, these requirements are that every aircraft in flight or moving on the ground during the hours of darkness shall display:

(a) A green light at or near the starboard wing tip, visible in the horizontal plane from a point directly ahead through an arc of 110 degrees to starboard.
(b) A red light at or near the port wing tip, with a similar arc of visibility to port.
(c) A white light visible from the rear of the aircraft in the horizontal plane through an arc of 140 degrees. The conventional location of this light is in the aircraft's tail, but in certain cases, notably such aircraft as the Douglas DC-10 and Lockheed 1011 "Tristar", white lights are mounted in the trailing edge sections of each wing tip.

The above angular settings are indicated in Fig. 10.1.

The construction of the light fittings themselves varies in order to meet the installation requirements for different types of aircraft. In general, however, they consist of a filament type lamp, appropriate fitting and transparent coloured screen or cap. The screen is specially shaped and, together with the method of arranging the filament of the lamp, a sharp cut-off of light at the required angle of visibility is obtained. The electrical power required for the lights is normally 28 volts d.c. but in several current types of "all a.c." aircraft, the lights are supplied with 28 volts a.c. via a step-down transformer. The operation of navigation lights, and their circuit arrangements, are factors which are dictated primarily by the regulations established for the flight operation of the types of aircraft concerned. Originally lights were required to give steady lighting conditions, but in order to improve the position marking function, subsequent

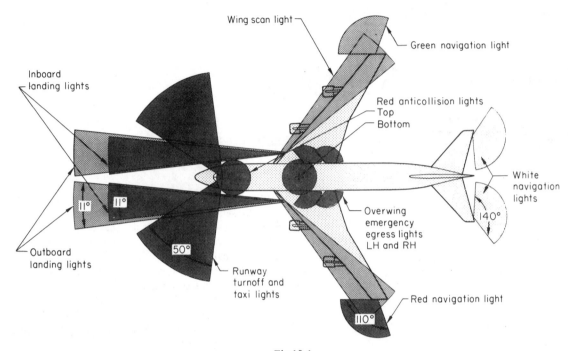

Fig 10.1
Disposition of external lighting

developments provided for the lights to flash in a controlled sequence. However, following the adoption of flashing anti-collision beacons the requirement for flashing navigation lights was discontinued and the requirement for steady lighting conditions re-introduced to become the order of the day once more. It is possible, however, that flashing navigation lights may still be observed on occasions; these are installed in some aircraft below a certain weight category, registered before current requirements became effective and thereby permitted alternative lighting arrangements.

ANTI-COLLISION LIGHTING

Anti-collision lighting also fulfils a position marking function and, in conjunction with navigation lights giving steady lighting conditions, permits the position of an aircraft to be more readily determined. A lighting system may be of the type which emits a rotating beam of light, or of the strobe type from which short-duration flashes of high-intensity light are emitted. In some current types of aircraft both methods are used in combination, the strobe lighting forming what is termed "supplementary lighting".

Rotating Beam Lights. These lights or beacons as they are often called, consist of a filament lamp unit and a motor, which in some cases drives a reflector and in others the lamp unit itself; the drive transmission system is usually of the gear and pinion type and of a specific reduction ratio. All components are contained within a mounting enclosed by a red glass cover. The power required for beacon operation is normally 28-volts d.c., but a number of types are designed for operation from an a.c. supply, the motor requiring 115 volts and the lamp unit 28 volts supplied via a step-down transformer. The motor speed and gear drive ratios of beacons are such that the reflector or lamp unit, as the case may be, is operated to establish a beam of light which rotates at a constant frequency. Typical speeds are 40–45 rev/min giving a frequency of 80–90 cycles per minute. There are several variations in the design of beacons, but the two types described here usefully serve as examples of how the rotating reflector and rotating lamp techniques are applied.

The beacon shown in Fig. 10.2 employs a V-shaped reflector which is rotated at about 45 rev/min by a d.c. motor, over and about the axis of a sealed beam lamp. One half of the reflector is flat and emits a narrow high-intensity beam of light near the horizontal, while the other half is curved to increase the up and down spread of its emitted beam to 30 degrees above and below the horizontal, and thereby reducing the light intensity.

Figure 10.3 illustrates a beacon employing two filament lamps mounted in tandem and pivoted on their own axes. One half of each lamp forms a reflector, and the drive from the motor is so arranged that the lamps oscillate through 180 degrees, and as may be seen from the inset diagram, the light beams are 180 degrees apart at any instant. The power supply required for operation is a.c.

Strobe Lighting. This type of lighting system is based on the principle of a capacitor-discharge flash tube. Depending on the size of the aircraft, strobe lighting may be installed in the wing tips to supplement the conventional red beacons, they may be used to function solely as beacons, or may be used in combination as a complete strobe type anti-collision high-intensity lighting system.

The light unit takes the form of a quartz or glass tube filled with Xenon gas, and this is connected to a power supply unit made up essentially of a capacitor, and which converts input power of 28 volts d.c. or

115 volts a.c. as the case may be, into a high d.c. output, usually 450 volts. The capacitor is charged to this voltage and periodically discharged between two electrodes in the Xenon-filled tube, the energy producing an effective high-intensity flash of light having a characteristic blue-white colour. A typical flashing frequency is 70 per minute.

The unit shown in Fig. 10.4 is designed for wing tip mounting and consists of a housing containing the power supply circuitry, the tube, reflector and glass lens. When used as supplementary lighting or as a complete strobe anti-collision lighting system, three units are installed in trailing positions in each wing tip, and all lights are controlled in a flashing sequence by controllers and flasher timing units.

LANDING LIGHTS AND TAXI LIGHTS
As their names indicate these lights provide essential illumination for the landing of an aircraft and for taxi-ing it to and from runways and terminal areas at night and at other times when visibility conditions are poor. Landing lights are so arranged that they illuminate the runway immediately ahead of the aircraft from such positions as wing leading edges, front fuselage sections and nose landing gear structure. The lights are of the sealed beam type and in some aircraft are mounted to direct beams of light at pre-

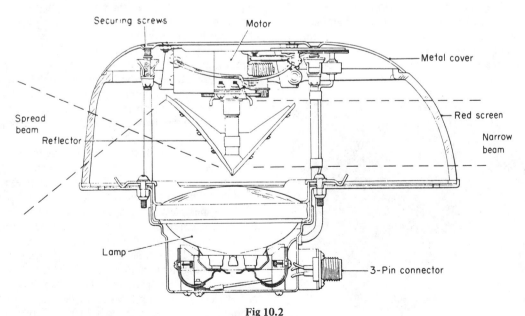

Fig 10.2
Rotating reflector beacon

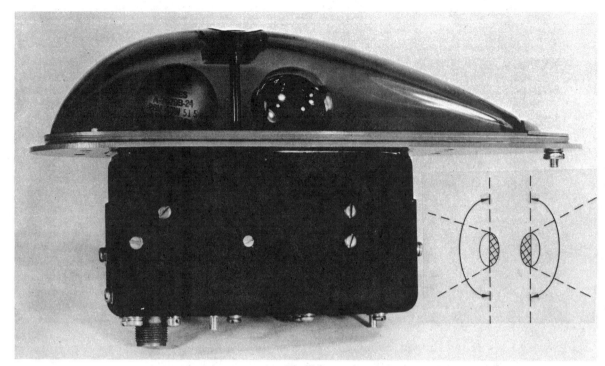

Fig 10.3
Rotating lamp beacon

determined and fixed angles. In other types of aircraft, the lights may be extended to preselected angles, and retracted, by an electric motor and gear mechanism, or by a linear actuator. Micro-type limit switches are incorporated in the motor circuit and are actuated at the extreme limits of travel to interrupt motor operation.

Fig 10.4
Typical strobe light unit

A typical power rating for lights is 600 watts, and depending on the design the power supply required for operation may be either d.c. or a.c. at 28 volts, the latter being derived from a 115-volts supply via a step-down transformer. In lights of the retractable type which require a.c. for their operation, the motor is driven directly from the 115-volts supply. The supplies to the light and motor are controlled by switches on the appropriate control panel in the cockpit. An example of a retractable type landing light is shown in Fig. 10.5.

The circuit of an extending/retracting light system is shown in Fig. 10.6. It is drawn to indicate the retracted position, and so the "retract" and "extend" limit switches controlling the motor, are open and closed respectively. The supply circuit to the light itself is automatically interrupted when it is retracted. When the control switch is placed in the "extend" position, the 115-volt supply passes through the corresponding field winding of the motor until interrupted by the opening of the extend limit switch. The retract limit switch closes soon after the motor starts extending the light. The switch in the supply circuit to the light also closes but the light is

not illuminated until it is fully extended and the
control switch placed in the "on" position. The
power supply to the light is reduced from 115 to
15 volts by a step-down transformer.

In some aircraft, a fixed-type landing light is
located in the leading edge of each wing near the
fuselage, and an extending/retracting type is located
in the fairing of each outboard landing flap track.
In lights located in flap track fairings, additional
switches are included in the "retract" and "extend"
circuits. The switches are actuated by a mechanical
coupling between the wing and flap track fairings.
Thus, when the landing flaps are lowered, and the
landing lights extended, the circuits of the motor will
be signalled to adjust the positions of the lights so
that their beams remain parallel to a known fore
and aft datum regardless of flap positions.

Taxi lights are also of the sealed beam type and are
located in the fuselage nose section, in most cases on
the nose landing gear assembly. The power rating of
the lights is normally lower than that of landing lights
(250 watts is typical) and the supply required is
either d.c. or a.c. at 28 volts.

In certain cases the function of a taxi light is
combined with that of a landing light. For example,
in the unit illustrated in Fig. 10.5, the light has two
filaments, one rated at 600 watts and the other at 400
watts; both filaments provide the illumination for
landing, while for taxi-ing only the 400 watt filament
is used.

Fig 10.5
Typical landing light

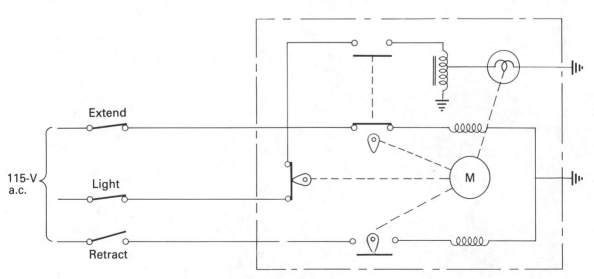

Fig 10.6
Extending/retracting light circuit

150

In addition to taxi lights some of the larger types of transport aircraft are equipped with lights which direct beams of light to the sides of the runway (see Fig. 10.1). These are known as runway turn-off lights, their primary function being to illuminate the points along the runway at which an aircraft must turn to leave the runway after landing.

ICE INSPECTION LIGHTS

Ice inspection or wing-scan lights are fitted to most types of transport aircraft, to detect the formation of ice on the leading edges of wings and also at the air intakes of turbine engines. Lights are also of the sealed beam d.c. or a.c. type and with power ratings varying from 60 watts to 250 watts depending on the lighting intensity required for a particular aircraft type. They are recessed into the sides of the fuselage and are preset to direct beams of light at the required angles. In some aircraft having rear-mounted engines lights are also recessed into the trailing edge sections of the wings.

Internal Lighting

The internal lighting of aircraft can be broadly divided into three categories: cockpit or operational lighting, passenger cabin lighting, and servicing lighting which includes galleys, toilet compartments, freight compartments and equipment bays.

COCKPIT LIGHTING

The most important requirements for cockpit lighting are those necessary to ensure adequate illumination of all instruments, switches, controls, etc., and of the panels to which these items are fitted. Some of the

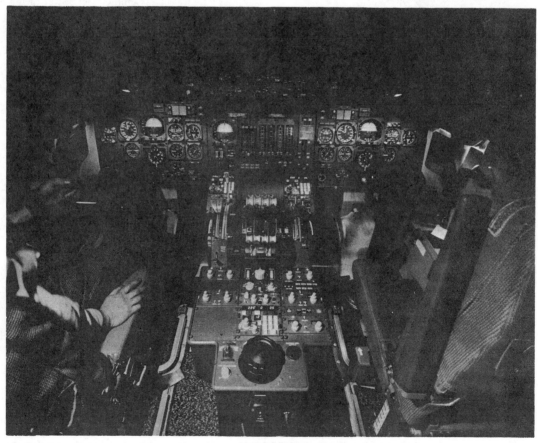

Fig 10.7
Boeing 747 cockpit under night lighting conditions

methods adopted to meet these requirements are as follows:

(i)   integral lighting, i.e. one in which the light source is within each instrument;

(ii) pillar and bridge lighting, in which a number of lights are positioned on panels to illuminate small adjacent areas, and to provide flood-lighting of individual instruments;

(iii) flood-lighting, whereby lamps are positioned around the cockpit to flood-light specific panels or a general area.

(iv) trans-illuminated panels which permit etched inscriptions related to various controls, notices and instructions to be read under night or poor visibility conditions.

A view of the Boeing 747 cockpit under night lighting conditions is shown in Fig. 10.7.

INTEGRAL LIGHTING

The principal form of integral lighting for instruments is that known as wedge or front lighting; a form deriving its name from the shape of the two portions which

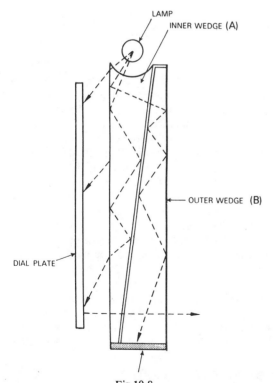

Fig 10.8
Wedge-type lighting

together make up the instrument cover glass. It relies for its operation upon the physical law that the angle at which light leaves a reflecting surface equals the angle at which it strikes that surface. The two wedges are mounted opposite to each other and with a narrow airspace separating them as shown in Fig. 10.8. Light is introduced into wedge "A" from two 6-volt lamps set into recesses in its wide end. A certain amount of light passes directly through this wedge and on to the face of the dial while the remainder is reflected back into the wedge by its polished surfaces. The angle at which the light rays strike the wedge surfaces governs the amount of light reflected; the lower the angle, the more light is reflected.

The double wedge mechanically changes the angle at which the light rays strike one of the reflecting surfaces of each wedge, thus distributing the light evenly across the dial and also limiting the amount of light given off by the instrument. Since the source of light is a radial one, the initial angle of some light rays with respect to the polished surfaces of wedge "A" is less than that of the others. The low-angle light rays progress further down the wedge before they leave and spread light across the entire dial. Light escaping into wedge "B" is confronted with constantly decreasing angles, and this has the effect of trapping the light within the wedge and directing it to its wide end. Absorption of light reflected into the wide end of wedge "B" is ensured by painting its outer part black.

PILLAR AND BRIDGE LIGHTING

Pillar lighting, so called after the method of construction and attachment of the lamp, provides illumination for individual instruments and controls on the various cockpit panels. A typical assembly, shown in Fig. 10.9, consists of a miniature centre-contact filament lamp inside a housing, which is a push fit into the body of the assembly. The body is threaded externally for attachment to the panel and has a hole running through its length to accommodate a cable which connects the positive supply to the centre contact. The circuit through the lamp is completed by a ground tag connected to the negative cable.

Light is distributed through a filter and an aperture in the lamp housing. The shape of the aperture distributes a sector of light which extends downwards over an arc of approximately 90 degrees to a depth slightly less than 2 in. from the mounting point.

The bridge-type of lighting (Fig. 10.9(b)) is a multi-lamp development of the individual pillar lamp

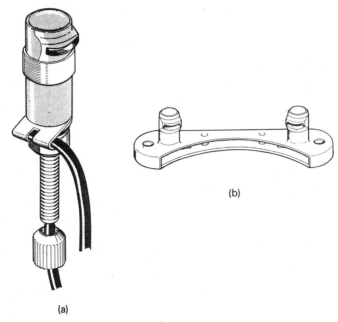

(b)

(a)

Fig 10.9
Pillar and bridge lighting

already described. Two or four lamps are fitted to a bridge structure designed to fit over a variety of the standardized instrument cases. The bridge fitting is composed of two light alloy pressings secured together by rivets and spacers, and carrying the requisite number of centre contact assemblies above which the lamp housings are mounted. Wiring arrangements provide for two separate supplies to the lamps thus ensuring that total loss of illumination cannot occur as a result of failure of one circuit.

These panels or "lightplates", provide for the illumination of system nomenclature, switch positions etc. They are of plastic through which light from many very small incandescent bulbs is passed. The light can only be seen where appropriate characters have been etched through a painted surface of a panel. The bulbs are soldered in place and are not replaceable when installed. More than one bulb provides illumination in each relevant area so that failure of a bulb will not impair illumination.

FLOOD-LIGHTING
Flood-lighting is used for the general illumination of instruments, control panels, pedestals, side consoles and areas of cockpit floors. The lights usually take the form of incandescent lamp units and fluorescent

tube units and depending on the type of aircraft, both forms may be used in combination.

ELECTROLUMINESCENT LIGHTING
This form of lighting is employed in a number of aircraft as passenger information signs and also, in some cases, for the illumination of instrument dials and selective positions of valves or switches. An electroluminescent light consists of a thin laminate structure in which a layer of phosphor is sandwiched between two electrodes, one of which is transparent. The light requires a.c. for its operation, and when this is applied to the electrodes the phosphor particles luminesce, i.e. visible light is emitted through the transparent electrode. The luminescent intensity depends on the voltage and frequency of the a.c. supply. The area of the phosphor layer which becomes "electroluminescent" when the current is applied is that actually sandwiched between the electrodes; consequently if the back electrode is shaped in the form of a letter or a figure the pattern of light emitted through the transparent electrode is an image of the back electrode.

PASSENGER CABIN LIGHTING
The extent to which lighting is used in a passenger cabin depends on the size of a cabin and largely on

the interior decor adopted for the type of aircraft; thus, it can vary from a small number of roof-mounted incandescent lamp fittings to a large number of fluorescent fittings located in ceilings and hat racks so as to give concealed, pleasing and functional lighting effects. The power supplies required are d.c. or a.c. as appropriate, and in all commercial passenger transport aircraft the lights are controlled from panels at cabin attendant stations. In addition to main cabin lighting, lights are also provided for passenger service panels (see p. 182) and are required for the illumination of essential passenger information signs, e.g. "Fasten Seat Belts" and "Return to Cabin". The lights for these signs may be of the incandescent type or, in a number of aircraft, of the electroluminescent type described earlier. They are controlled by switches on a cockpit overhead panel.

Control of Lighting Intensity

Certain internal lighting circuits must have a means of varying the light intensity and so they are provided with an intensity control system. The methods of control, and their application, depends largely on the extent of the lighting required, this in turn, being dependent on the type of aircraft. The fundamental operating principles of each method are shown in Fig. 10.10.

The most basic of dimming circuits is the one utilizing a panel-mounted rheostat which is connected in series with the lights whose intensity is to be controlled (diagram (a)). Power from the d.c. busbar is fed to the rheostat wiper which, at contact position "A" isolates the lights from the supply. When moved to contact position "B", the circuit is switched on but as current must flow through the whole of the rheostat resistance, the lights will be dimly illuminated. As the wiper is moved towards contact position "C" the resistance in the circuit becomes less and less and so the lighting intensity increases. At position "C" maximum current flows through the circuit to provide maximum lighting intensity.

Diagram (b) illustrates a circuit development of the basic rheostat method and is one which is widely adopted in many aircraft since it permits the use of less "bulky" rheostats, and control of an increased number of lights in any one circuit. The circuit utilizes an NPN transistor which functions as a remotely controlled resistor unit. A rheostat is still required to vary the voltage input to the transistor,

but because a transistor requires only very low voltage levels over its conducting range, the rheostat can be smaller from the point of view of electrical characteristics and physical dimensions.

D.c. power is supplied to the rheostat and also to the collector "C" of the transistor. When the rheostat wiper is at contact position "A", the voltage at the

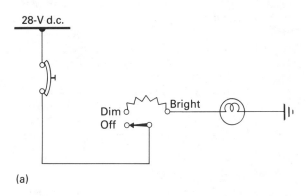

(a)

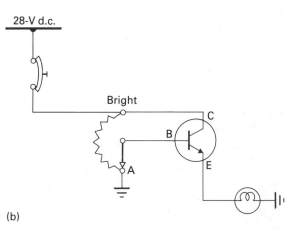

(b)

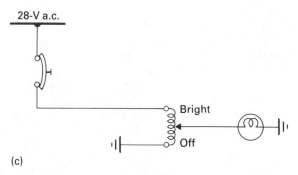

(c)

Fig 10.10
Control of lighting intensity

154

base of the transistor is zero, and no current flows through the collector to the emitter "E" or out to the lights. Movement of the wiper from contact position "A", causes a positive voltage to be applied to the base of the transistor, and a small amount of current flows from the collector, and through the emitter to the lights as a result of a reduction in resistance of the collector-emitter junction. Further movement of the wiper increases the positive voltage at the transistor base, and the resulting decrease in collector-emitter junction resistance increases the current flow to the lights and therefore, their intensity.

Diagram (c) of Fig. 10.10, shows a method in which lighting intensity may be controlled by means of a variable transformer. This is commonly adopted in aircraft whose main power generating systems are a.c.

EMERGENCY LIGHTING

An essential requirement concerning lighting is that adequate illumination of the cockpit and the various sections of the cabin, exits, escape hatches, chutes etc., must be provided under emergency conditions, e.g. a crash-landing at night. The illumination is normally at a lower level than that provided by the standard lighting systems, since the light units are directly powered from an emergency battery pack or direct from the aircraft battery in some cases. The batteries are normally of the nickel-cadmium type although in some aircraft silver-zinc batteries are employed.

Under normal operating conditions of the aircraft, emergency battery packs are maintained in a fully-charged condition by a trickle charge from the aircraft's main busbar system.

Primary control of the lights is by means of a switch on a cockpit overhead panel.

Engine Starting Systems

Throughout the development of aircraft engines a number of methods of starting them have been used and the prime movers involved have varied from a mechanic manually swinging a propeller, to electric motors and electric control of sophisticated turbo-starter units. Although there are still one or two types of light aircraft in service requiring the manual swinging technique, the most widely adopted starting method for reciprocating engines utilizes electric motors, while for the starting of gas turbine engines either electric motors or turbo-starter units may be utilized as the prime movers.

ELECTRIC STARTER MOTOR SYSTEMS

In basic form, these systems consist of a motor, an engaging gear, a relay and a starter switch; in some systems a clutch mechanism is also incorporated in the engaging gear mechanism. The motors employed may be of the plain series-field type or may be compounded with a strong series bias (see pp. 138 and 139).

Fig. 10.11 shows the interconnection of the principal electrical components typical of those required for the starting of reciprocating engines installed in many types of light aircraft. When the starter switch is closed, direct current from the battery and busbar energizes the starter relay, the closed contacts of which connect the motor to the battery. The relay contacts are of the heavy-duty type to carry the high current drawn by the motor during the period of cranking over the engine.

The method of engaging a motor with an engine varies according to the particular engine design. For most types of light aircraft engines, a pinion is engaged with a starter gear ring secured to the engine crankshaft in a manner similar to that employed for starting automobile engines. When the engine starts, it overruns the starter motor and the pinion gets "kicked out" of engagement. In other versions used for starting more powerful engines, a jaw engages with a similar member on the engine and the drive is transmitted via a clutch and reduction gear train in the starter motor and in an accessories gearbox in the engine.

The gear ratio between a starter motor and a reciprocating engine is such that it provides a low cranking speed of the engine; a typical reduction ratio is about 100 : 1. Cranking speed is not critical because of the fuel priming provisions made in the starting drill, and also because there is a good stream of sparks available at the plug points for the power stroke. Thus, once the engine has "fired" and gets away under its own power further assistance from the starter motor is rendered unnecessary. Although the moment of inertia of an engine's moving parts is comparatively light during cranking, a starter motor has to overcome some heavy frictional loads, i.e. loads of pistons and bearings, and also loads due to compression.

TURBINE ENGINE STARTING

Compared with a reciprocating engine, the starting of a turbine engine represents a relatively severe duty for the starter motor. This stems mainly from the starting principle involved and also from the construction of the rotating assembly, e.g. whether the com-

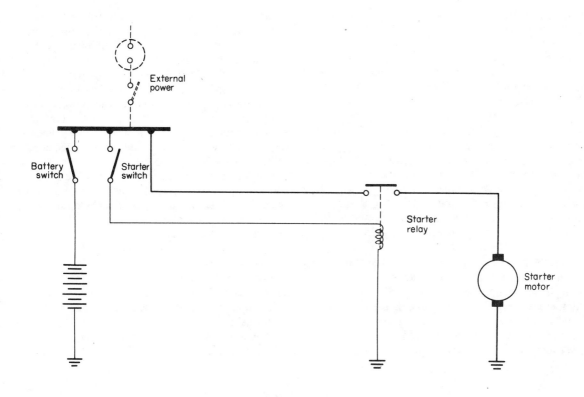

Fig 10.11
Simple engine starting system

pressor and turbine are on a single shaft (single-spool engine) or whether high-pressure compressor/turbine assemblies and low-pressure compressor/turbine assemblies on separate shafts are employed (two-spool engine). Another factor also is whether the compressor and turbine assembly is designed to drive a propeller. In general, turbine engines have a high moment of inertia, and since it is a requirement that starting shall be effected as quickly as possible, then high gear ratios and therefore high cranking speeds are necessary.

The process of starting a turbine engine involves the provision of an adequate and continuous volume of air to the combustion system, effective atomization of fuel at the burners of the combustion system, and the initiation of combustion in the combustion chambers. To provide the necessary volume of air the starter motor must be capable of developing sufficient power to accelerate the compressor smoothly

and gently from a static condition to a fairly high speed. At some stage in the cranking operation, fuel is injected into the combustion system and the fuel/air mixture is ignited, i.e. the engine "fires" or "lights-up" as it is more usually stated in turbine engine terms. Unlike reciprocating engine starting, however, the starter motor does not disengage at this point but, assisted by the engine, continues to accelerate it up to a speed at which the engine alone is capable of maintaining rotation. This is known as the self-sustaining speed of the engine. Eventually a condition is reached where the starter motor is no longer required and its torque, and the current consumed, start decreasing fairly rapidly. Its speed will tend to increase, but this is limited by the retarding torque provided by the shunt field when there is no longer a load on the motor (see also p. 138). Depending on the type of starter system, the power supply to the starter motor is interrupted automatically either by

156

the decrease in current causing the starter relay to de-energize, or by the opening of contacts in a time switch unit.

Figure 10.12 illustrates the circuit diagram of a system based on that employed in a current type of twin turbopropeller aircraft for the starting of its engines. The starter motor is a 28-volts d.c. four-pole compound-wound machine having a torque output of 16·5 lbf.ft(22·37 Newton metre) at a speed of 3800 rev/min and a time rating of 90 seconds. It drives the engine through a clutch, pawl mechanism and reduction gear. The clutch is held in the driving position until the engine has accelerated above the starter motor speed and until the centrifugal force acting on the pawl mechanism is sufficient to release the pawls. The starter motor is disengaged by the action of an overspeed relay.

When the master switch is set to the "start" position, and the starter push switch is depressed, direct current flows through the coil of the main starter relay thereby energizing it. At the same time current also flows to contact "1" of the overspeed relay. The closing of the heavy-duty contacts "A" and "B" of the starter relay completes a circuit from the main

busbar to the starter motor via the coil of the overspeed relay, which on being energized, allows current to flow across its contacts to the coil of the push switch thereby holding this switch closed. During initial stages of starting the current drawn by the starter motor is high, and as this is carried by the coil of the overspeed relay continued cranking of the engine is assured. As the engine accelerates, the starter motor draws less current until, at a value predetermined by the speed at which the engine becomes self-sustaining, the overspeed relay is de-energized, this in turn de-energizing the starter switch and main starter relay. The overspeed relay therefore prevents the starter motor from overspeeding by ensuring that the power supply is disconnected before the starter drive is disengaged from the engine.

The purpose of the "blow out" position of the master switch is to permit the engine to be cranked over in order to blow out unburnt fuel resulting from an unsuccessful start or "light up". When the position is selected, the circuit is operated in a similar manner to normal starting except that the starter switch must be pulled to the "off" position after the motor has been running for 30 seconds. The reason for this is

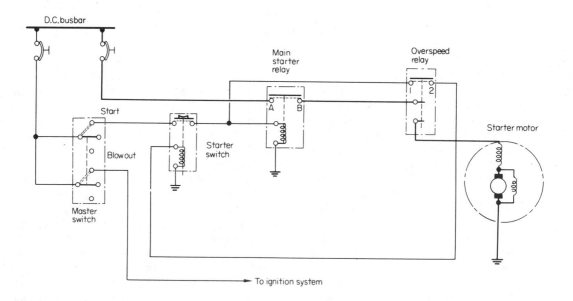

Fig 10.12
Basic circuit of a turboprop engine starting system

that since the ignition system is isolated, the starter motor is still heavily loaded and so the current through the overspeed relay remains too high for the relay to de-energize of its own accord.

TURBO-STARTER SYSTEMS

With the development of more powerful turbine engines ever-increasing power output from starter systems was required for effective starting action. As far as electrical methods of starting were concerned this presented increasingly difficult problems associated notably with high current demand, increased size and weight of motors and cables. These problems therefore led to the discontinuance of electric motors for the starting of powerful engines, and their functions were taken over by turbo-starter systems requiring a simpler control circuit consuming only a few amperes.

There are three principal types of turbo-starter systems; air, cartridge and monofuel, the application of each being governed largely by the operational role of the aircraft, i.e. civil or military. The basic principle is the same for each system, that is, a gas is made to impinge on the blades of a turbine rotor within the starter unit, thereby producing the power required to turn the engine shaft via an appropriate form of coupling.

The gas may be (i) compressed air supplied to a turbine air motor from either an external supply unit, an A.P.U. in the aircraft or the compressor of a running engine; (ii) the cordite discharge from an electrically fired cartridge or (iii) the result of igniting a monofuel, in other words a fuel which burns freely without an oxidant such as air; a typical fuel is isopropylnitrate.

The electrical control circuits normally require d.c. for their operation, their function being to energize solenoid-operated air control valves, to fire cartridge units and to energize a fuel pump motor and ignition systems as appropriate to the type of turbo-starter unit.

STARTER-GENERATOR SYSTEMS

Several types of turbine-powered aircraft are equipped with starter systems which utilize a prime mover having the dual function of engine starting and of supplying power to the aircraft's electrical system. Starter-generator units are basically compound-wound machines with compensating windings and interpoles, and are permanently coupled with the appropriate engine via a drive shaft and gear train. For starting

purposes, the unit functions as a fully compounded motor, the shunt winding being supplied with current via a field changeover relay. When the engine reaches self-sustaining speed and the starter motor circuit is isolated from the power supply, the changeover relay is also automatically de-energized and its contacts connect the shunt-field winding to a voltage regulator. The relay contacts also permit d.c. to flow through the shunt winding to provide initial excitation of the field. Thus, the machine functions as a conventional d.c. generator, its output being connected to the busbar on reaching the regulated level.

Ignition Systems

All types of aircraft engines are dependent on electrical ignition systems. In reciprocating-type engines, the charges of fuel vapour and air which are induced and compressed in the cylinders, are ignited through the medium of sparks produced by electric discharges across the gaps between the electrodes of a spark plug fitted in each cylinder, and a continuous series of high-voltage electrical impulses, separated by intervals which are related to engine speed, must be made available to each of the plugs throughout the period the engine is running. A basically similar electrical ignition system is also used to initiate combustion of the fuel/air mixture in the combustion chambers of gas turbine engines. It is, however, of much simpler form for the reasons that impulse intervals are not related to engine speed, and as combustion is continuous after "light up", the ignition system is only required during the starting period.

Reciprocating-type engine ignition systems fall into one or other of two main categories; coil ignition and magneto ignition. The former derives its power from an external source, e.g. the main power supply, while the magneto is a self-contained unit driven by the engine and supplying power from its own generator. In aircraft engine applications, magneto ignition is the system most commonly adopted.

MAGNETO IGNITION SYSTEMS

Magneto ignition systems, which operate on the principles of electromagnetic induction, are classified as either high tension or low tension, and they consist of the principal components shown schematically in Fig. 10.13. Most of these components are contained within the magneto, which is basically a combination of permanent-magnet a.c. generator and autotransformer.

The high tension system is the one most widely used, and the requisite alternating fluxes and voltages are induced either by rotating the transformer windings between the poles of a permanent magnet, by rotating the magnet between fixed transformer windings or by rotating soft-iron inductor bars between fixed permanent magnet and transformer windings. These arrangements, respectively, permit further classification of magnetos as (i) rotating armature, (ii) rotating magnet and (iii) polar inductor.

The rotating portion of a magneto is driven by the engine through a coupling and an accessory gear drive shaft. As the windings are cut by the alternating magnetic flux from the appropriate source, a low voltage is induced in the primary winding to produce a current and flux of a strength directly proportional to the rate at which the main flux is cut. At this point the primary circuit is broken by the contact breaker, the contacts, or points, of which are opened by a cam driven by the rotating assembly. The primary flux therefore collapses about the secondary winding, which produces a high voltage output. The output is, however, not sufficient to produce the required discharge at the spark plugs and it is necessary to speed up the rate of flux collapse. This is effected by connecting a

capacitor across the contact breaker so that the capacitor is shorted out when the breaker points are closed and is charged by primary winding current when the points are open. When the potential difference across the capacitor reaches the point whereby it discharges itself, the correspondingly high current flows through the primary winding in the reverse direction and thereby rapidly suppresses the primary flux to produce the required higher secondary output voltage. In addition to this function, the capacitor also prevents arcing between the contact breaker points as they begin to open, thereby preventing rapid deterioration of the points.

The secondary winding output is supplied to the distributor, the purpose of which is to ensure that the high voltage impulses are conducted to the sparking plugs in accordance with the order in which combustion must take place in each cylinder, i.e. the "firing" order of the engine. A distributor consists of two main parts, a rotor made up of an insulating and a conducting material, and a block of insulating material containing conducting segments corresponding in number to the number of cylinders on the engine. The conducting segments are located circumferentially around the distributor block in the desired firing order, so that as the rotor turns a circuit is

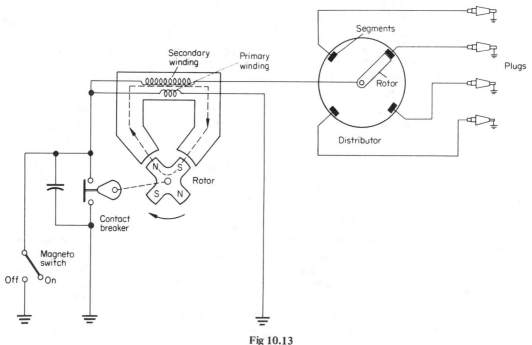

Fig 10.13
Magneto ignition system

completed to a sparking plug each time there is alignment between the rotor and a segment.

Distributors usually form part of magnetos, and the rotors are rotated at the required speed by a gear driven from the main magneto shaft. In some cases, however, distributors may be separate units driven by an engine gear train and drive shaft. To prevent ionization, and to minimize "flashover", the distributor casing is vented to atmosphere, and in many types of magnet a flameproof wire mesh screen is provided to prevent combustion of any flammable vapours round the engine.

MAGNETO AND DISTRIBUTOR SPEEDS

Ignition of the combustible mixture is required in each cylinder once in every two revolutions of the engine crankshaft, and as a result there must be a definitie relationship between such factors as the number of sparks produced by a magneto and the speeds of the magneto, distributor and engine. Magneto speed may be calculated from the relation:

$$\frac{number\ of\ cylinders}{2 \times magneto\ sparks\ per\ rev.}$$

A rotating armature magneto, which is normally only used on engines having up to six cylinders, produces two sparks per rev. Thus, assuming that one is fitted to a four-cylinder engine then it must be driven at the same speed as the engine. A rotating magnet or polar inductor magneto produces four sparks per rev and is normally used on engines having more than six cylinders. Thus, for a twelve-cylinder engine the magneto must be driven at one and a half times the engine speed. Distributor rotors are driven at half engine speed irrespective of magneto speed.

AUXILIARY STARTING DEVICES

As mentioned earlier, during starting, a piston engine is cranked over at very low speeds, and as a result its magnetos are driven much too slowly for the e.m.f. induced in the primary winding to produce a spark of adequate energy-content at the instant the contact breaker points open. It is therefore necessary to provide auxiliary means for boosting the magneto output during the engine starting period, when it is advantageous to have the spark retarded to some extent. Two methods widely adopted are impulse starters and booster coils, which are described in the following paragraphs. The retarding of the spark is effected by a secondary brush in the distributor arm which "trails" the main brush.

IMPULSE COUPLINGS

Impulse couplings, or impulse starters as they are sometimes called, are used in some small piston engine ignition systems and are fitted between the magneto shaft and drive shaft. The unit produces a heavy spark by giving a magneto armature or magnet a brief acceleration at the moment of spark production. In one type of unit the coupling between the magneto and engine is a spring-loaded clutch device which flicks the armature or magnet through the positions at which a spark normally occurs, thus momentarily increasing its rotational speed and the voltage generated. After the engine is started and the magneto reaches a speed at which it furnishes sufficient output, flyweights in the coupling fly outward due to centrifugal force and overcome the springs, so that the coupling functions as a solid drive shaft and the magneto continues to operate in the normal manner.

BOOSTER COILS

Booster coils, which may be either of the high tension impulse or low tension impulse type, derive their power from the aircraft's system via either the battery or the ground power supply source. The supply is controlled either by a separate booster coil or the engine starter switch. High tension booster coils supply a stream of impulses to the trailing brush of the distributor, while in a low tension system a stream of impulses is fed to the magneto primary windings either to augment or to replace the voltage induced by the magnetic flux. In some low tension systems, the supply to the primary winding is fed via a second contact-breaker, which is retarded in relation to the main contact-breaker but connected in parallel with it. With this arrangement intermittent high tension current is induced in the secondary winding of the magneto.

IGNITION SWITCHES

Ignition systems are controlled by "on-off" switches connected in the magneto circuit, but unlike the basic and conventional switching arrangements, an ignition system switch completes a circuit by closing its contacts in the "off" position. The circuit in this case is between the magneto primary winding and ground, and since the contact-breaker becomes short-circuited, then in the event the magneto is rotated, there can be no sudden collapse of the primary winding flux and therefore no high voltage spark across the spark plug gap.

On dual-ignition systems each magneto may be controlled by a separate toggle switch or, as is more usual, by a rotary type four-position switch controlling both magnetos. The four positions are "off", "left", "right" and "both". The left and right positions allow one system to be turned off at a time for carrying out "magneto-drop" checks during engine ground running.

LOW TENSION MAGNETO SYSTEMS

These systems were developed for use on engines having a large number of cylinders and designed for high altitude operation. They overcome certain problems which can occur with high tension systems, e.g. breakdown of insulation within a magneto due to decreased atmospheric pressure and electrical leakage, particularly when ignition harnesses are enclosed in metal conduits. Furthermore, the amount of cable carrying high voltages is considerably reduced. The magneto is similar to a polar inductor type of magneto but does not embody a secondary coil. Low voltage impulses from the magneto primary winding are supplied directly to the distributor, which also differs from the types normally employed, in that voltage impulses are received and distributed via a set of brushes and segmented tracks. The distributor output is supplied to transformers corresponding in number to the number of spark plugs and located near the plugs. Thus, high voltage is present in only short lengths of shielded cable. Low tension magnetos are switched on and off in the same manner as high tension magnetos.

SPARK PLUGS

The function of a spark plug is to conduct the high voltage impulses from the magneto and to provide an air gap across which the impulses can produce a spark discharge to ignite the fuel/air charge within the cylinder.

The types of spark plugs used vary in respect to heat range, thread size, or other characteristics of the installation requirements of different engines, but in general they consist of three main components: (i) outer shell, (ii) insulator and (iii) centre electrode. The outer shell, threaded to fit into the cylinder, is usually made of high tensile steel and is often plated to prevent corrosion from engine gases and possible thread seizure. The threads are of close tolerance and together with a copper washer they prevent the very high gas pressure escaping from the cylinder. Pressure that might escape through the plug is retained by inner seals between the outer shell, the insulator, and centre electrode assembly.

The materials used for insulators vary between plug designs and applications to specific engines; those most commonly adopted are mica, ceramic and aluminium oxide ceramic, the latter being specifically developed to withstand more exacting mechanical, thermal and electrical requirements. Insulation is also extended into a screen tube which is fixed to the outer shell and provides attachment for the ignition harness cable to ensure suppression of radio interference.

The centre electrode carries the high tension voltage from the distributor and is so secured that the requisite spark gap is formed between it and a negative or ground electrode secured to the "firing" end of the outer shell. Electrodes must operate under very severe environmental conditions, and the materials normally chosen are nickel, platinum and iridium.

DUAL IGNITION

Almost all piston engines employ two entirely independent ignition systems; thus each cylinder has two spark plugs, each supplied from a different magneto. The purpose of dual ignition is to (i) reduce the possibility of engine failure because of an engine fault and (ii) reduce the time taken to burn the full charge enabling peak gas pressure to be reached and thereby increasing engine power output. Both magnetos are normally switched by a rotary switch in the manner already described.

TURBINE ENGINE IGNITION SYSTEMS

The ignition system of a turbine engine is much simpler than that of a piston engine due to the fact that fewer components are required and that electrical ignition of the air/fuel mixture is only necessary when starting an engine. Another difference is that the electrical energy developed by the system is very much higher in order to ensure ignition of atomized fuel under varying atmospheric and air mass flow conditions and to meet the problems of relighting an engine in the air.

The principal components of a system are a high-energy ignition unit and an igniter plug interconnected as shown in Fig. 10.14. Two such systems are normally fitted to an engine, the igniter plugs being located in diametrically-opposed combustion chambers to ensure a positive and balanced light-up during starting. Direct current from the aircraft's main busbar is supplied to an induction coil or a transistorized high tension generator within the ignition unit in conjunction with

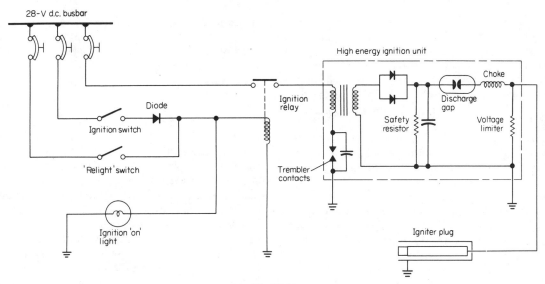

Fig 10.14
High-energy ignition system

the starter system, and also independently through the "relight" circuit. The coil, or generator, as appropriate, repeatedly charges a reservoir capacitor until its voltage, usually of the order of 2,000 volts, is sufficient to break down the sealed discharge gap. The gap is formed by two tungsten electrodes within a chamber exhausted of air, filled with an inert gas and sealed to prevent oxidation which would otherwise occur with the large current handled.

The discharge is conducted through a choke, which extends the duration of the discharge, and through a high tension lead to the igniter plug (see Fig. 10.15) at which the energy is released. A pellet at the "firing" end of the plug has a semi-conducting surface, and during operation this permits a minute electrical leakage from the centre electrode to the body, thereby heating the surface. Due to the negative temperature/resistance characteristics of the pellet a low resistance path is provided for the energy, which discharges across the surface as a high intensity flashover as opposed to a spark jumping an air gap. The capacitor recharges and the cycle is repeated approximately once every second. Once the fuel/air mixture has been ignited, the flame spreads rapidly through balance pipes which interconnect all the combustion chambers; thus combustion is self-sustaining and the ignition system can be switched off. The energy stored in the capacitors is potentially lethal, and to ensure their discharge when the d.c. supply is disconnected, the output is connected to ground via a safety resistor.

The electrical energy supplied by the ignition unit is measured in joules, and independent ignition systems normally consist of two units rated at 12 joules each.

In the event that through adverse flight conditions the flame is extinguished, the engine is "relit" by switching on the ignition system until the engine runs normally again. During relighting it is unnecessary to use the starter motor since the engine continues to rotate under the action of "windmilling". In some

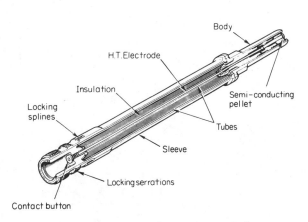

Fig 10.15
High-energy igniter plug

cases, relighting is automatic by having one of the two ignition units of a low rating (usually 3 joules) and keeping it in continuous operation. Where this method is not desirable a glow plug is sometimes fitted in the combustion chamber where it is heated by the combustion process and remains incandescent for a sufficient period of time to ensure automatic re-ignition.

In some types of aircraft the high energy ignition units are dependent on an initial power supply of 115 volts a.c., and the simplified circuit diagram of one such unit employed on the Boeing 747, is illustrated in Fig. 10.16.

The a.c. power supply is applied to the primary winding of a step-up power transformer T1, via a radio noise filter network consisting of inductor L1 and a capacitor C1. The high voltage induced in the secondary winding of T1 is then rectified by the diodes 1 and 2, the current passing through them being limited by resistors R1 and R2. The rectified output charges the capacitor C2 until the stored voltage reaches the ionization potential of the discharge gap. The discharge flows through the primary winding of the high-tension auto-transformer T2, and is further boosted by a charge developed across capacitor C4. The voltage induced in the secondary winding is then of sufficient potential to provide the requisite discharge flashover across the igniter plug gap. Resistors R3 and R4 provide the means of dissipating the energy of the circuit in the event that the output of the igniter unit is open-circuited. In addition, they serve to "bleed-off"

any residual charge on capacitor C4 between successive flashovers, and so provide a constant level of triggering voltage from the secondary winding of transformer T2.

Fire Detection and Extinguishing Systems

Fire is, of course, one of the most dangerous threats to an aircraft and so precautions must be taken to reduce the hazard, not only by the proper choice of materials and location of equipment in potential fire zones, but also by the provision of adequate fire detection and extinguishing systems. These systems may be broadly classified as (i) fixed, some examples of which are used mainly for engine fire protection, and detection of smoke in baggage compartments, or (ii) portable, for use in the event of cabin fires. Both systems are employed in all aircraft except certain small low-powered piston engined types which, having been certificated as constituting a negligible fire risk, at most need only a portable extinguisher within the cockpit. Fixed detection and extinguishing systems only, require electrical power for their operation, and some typical examples are described in the following paragraphs.

FIRE DETECTION

A fire detection system is installed mainly in engine compartments, and consists of special detecting elements strategically positioned within several fire zones designated by the aircraft manufacturer. The elements, which may be of the "unit" or "spot" type

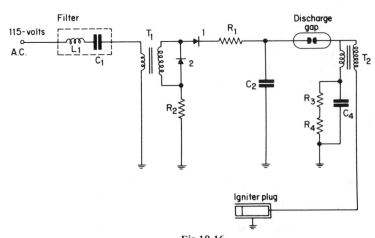

Fig 10.16
High-energy system (a.c. powered)

or the "continuous" wire type, are connected to warning lights and/or bells, and either type may be used separately, or together in a combined fire warning and engine overheat system.

Unit type detectors are situated at points most likely to be affected by fire, e.g. in an engine·breather outlet pipe, and the one most often used in engine compartments is of the differential expansion switch type, the principle of which was described on p. 106. These detectors may also be used for sensing an overheat condition in areas of the airframe structure adjacent to ducting supplying hot air for air-conditioning or de-icing systems.

In order to provide maximum coverage of an engine fire zone and to eliminate the use of a considerable number of unit detectors, a continuous wire type detector system (known as a "firewire" system) is normally used. The elements of a typical system take the form of various lengths of wire embedded in a temperature sensitive material within a small bore stainless steel or Inconel tube, and joined together by special coupling units to form a loop which may be routed round the fire zone as required. The wire and tube form centre and outer electrodes respectively and are connected to the aircraft's power supply via a control unit. The power supply requirements are 28 volts d.c. and 115 volts a.c. or, in some

systems, 28 volts d.c. only. Depending on the type of control unit the method of operation may be based on either variations in resistance or variations in capacitance with variations in temperature of the element filling material.

The electrical interconnection of components normally comprising a system is shown in Fig. 10.17. the control unit in this case is of the type employed with a variable resistance system. The a.c. supply is fed to a step-down transformer, while d.c. is supplied to the warning circuit via the contacts of a warning relay, the coil of which may be energized by the rectified output from the transformer secondary. With the test switch in the normal position, the ends of the centre wire electrode of the element are connected in parallel to the rectifier and to one end of the transformer secondary winding. The other end of the winding is connected to the outer tube or electrode so that the current path is always through the filling material, the resistance of which will govern the strength of rectified current flowing through the relay coil. With this arrangement the warning function is in no way affected in the event that a break should occur in the loop.

Under normal ambient temperature conditions the resistance of the filling material is such that only a small standing current flows through the material; therefore, the current flowing through the warning

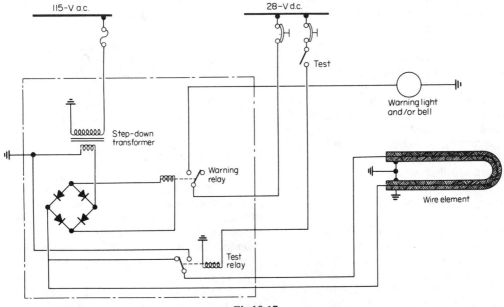

Fig 10.17
Fire detection system

relay coil is insufficient to energize it. In the event of a temperature rise the resistance of the filling material will fall since it has an inverse characteristic, hence the rectified current through the relay coil will increase, and when the fire zone temperature has risen to such a value that the relay coil current is at a predetermined level, it will energize the relay thereby completing the warning light or bell circuit. When the temperature falls and the current drops to a predetermined level the relay de-energizes and the system is automatically reset.

In a capacitance system the detector element is similar in construction to that earlier described, but in conjunction with a different type of control unit it functions as a variable capacitance system, the capacity of the element increasing as the ambient temperature increases. The element is polarized by the application of half-wave rectified a.c. from the control unit, which it stores and then discharges as a feedback current to the gate of a silicon-controlled rectifier (SCR) in the control unit during the non-charging half cycles. When the fire zone temperature rises the feedback current rises until at a pre-determined level the SCR is triggered to energize a fire warning light, or bell, relay. A principal advantage of this sytem is that a short circuit grounding the element or system wiring does not result in a false fire warning.

When the test switch is set to the "Test" position, the test relay is energized and its contacts change over the supply from the rectifier so that the current passes directly along the centre electrode. Thus, if there is no break in the loop there is minimum resisttance and the warning relay circuit is actuated to simulate a fire warning and so indicate continuity.

In some engine fire detection systems, detection is effected by two distinct sensing element loops; an "overheat" loop and a "fire" loop. An example of one such application based on the Lindberg Systron Donner system, is shown schematically in Fig. 10.18. This system, unlike that of the "firewire" system described earlier, utilizes sensing elements which trigger the warning circuits as a result of the temperature effects on the pressure of a gas.

An element consists of a stainless steel tube protected throughout its length by a teflon coating. Inside the tube is a metal hydride-coated element surrounded by an inert gas (helium). One end of the type is bifurcated and joined to two diaphragm-operated pressure switches, one in the open position and the other always kept closed by the normal pressure (20 psi) of the helium acting on its

diaphragm. The power required for system operation is 28 volts d.c. from the aircraft's battery busbar, and is supplied to the open switch contacts via those of the normally closed switch. Because the overall system is designed to sense two levels of temperature, then there must be two detecting elements each with a different temperature sensing level.

In the practical case, there are two pairs of elements connected as shown in Fig. 10.18, one pair forming the "overheat" loop, and the other pair the "fire" loop. The elements are located at the bottom of an engine, and on the engine side of the firewall.

If the local temperature rises to $205 \pm 30\,°C$, the coating of the element inside an "overheat" detector will release a gas (principally hydrogen). This will increase the pressure inside the tube so that the diaphragm of the normally-open switch will be displaced and its contacts closed. The signal now flowing from the contacts passes through an operational amplifier the output of which biases a transistor to allow the standing 28 volt d.c. to energize a relay. The closed contacts of the relay then complete a circuit to an amber "overheat" light, and also a light on a master caution panel.

If the local temperature should continue to increase and reach $315 \pm 30\,°C$, the pressure of the released hydrogen in a detecting element of the "fire" loop will trigger a signal to pass through a circuit similar to that of the overheat loop but, in this case, to illuminate red warning lights, and to set off an alarm bell.

In the event that the pressure of the helium inside a detector element should decrease, the normally-closed pressure switch would open to cause a "detector inoperative" light (not shown) to illuminate. Test circuits are therefore provided in the system so that the integrity of the normally-closed pressure switches can be checked.

SMOKE DETECTORS

In many of the larger types of transport aircraft, the freight holds, baggage compartments and equipment bays are often fitted with equipment designed for the detection of smoke. Detection equipment varies in construction, but in most cases the operation is based on the principle whereby air is sampled and any smoke present, causes a change of electric current within the detector circuit to trigger a warning system.

Figure 10.19 is a schematic arrangement of a smoke detector in use in some types of transport aircraft. The principal detecting elements are a

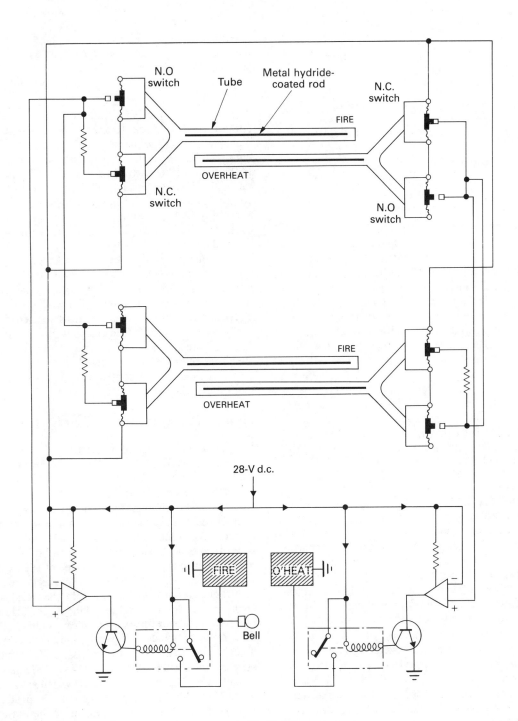

Fig 10.18
Overheat and fire detection system

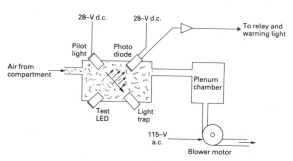

Fig 10.19
Smoke detector operation

pilot light, a light trap and a photo diode, disposed in a compartment or chamber as shown. The pilot light and photo diode are powered by 28-volt d.c.

Sampling tubes connect with the detector and plenum chambers and a blower motor powered by 115-volt a.c. When the system is armed, the blower motor draws air through the detector from the compartment in which the detector is located. The pilot light directs a beam to the light trap. If smoke is present to a level of 10 per cent, the light reflected from it as it passes through the beam will be detected by the photo diode. The current generated by the diode is then amplified to trigger a relay the contacts of which complete a circuit to the appropriate fire warning light. The light emitting diode (LED) forms part of a test circuit which activates the photo diode to simulate a smoke condition.

FIRE EXTINGUISHING

Fixed fire extinguishing systems are used mainly for the protection of engine installations, auxiliary power units, landing gear wheel bays and baggage compartments, and are designed to dilute the atmosphere of the appropriate compartments with an inert agent that will not support combustion. Typical extinguishing agents are methyl bromide, bromochlorodifluoromethane, freon, or halco, and these are contained within metal cylinders or "bottles" of a specified capacity. The agents are pressurized by an inert gas, usually dry nitrogen, the pressures varying between types of extinguisher, e.g. 250 lbf/in^2 for 12 pounds of methyl bromide, 600 lbf/in^2 for 4 pounds of freon. Explosive cartridge units which are fired electrically, are connected to distributor pipes and spray rings, or nozzles, located in the potential fire zones. Electrical power for cartridge unit operation is 28 volts d.c. and is supplied from an essential services busbar; the circuits are controlled by switches located in the cockpit. When the cartridge unit is fired a diaphragm is ruptured and the appropriate extinguishing agent is discharged through the distributor pipes and spray rings.

In the fire extinguisher systems of some types of aircraft, electrical indicators are provided to show when an extinguisher has been fired. An indicator consists of a special type of fuse and holder connected in the extinguisher cartridge unit circuit. The fuse takes the form of a small match-head type charge covered by a red powder and sealed within the fuse body by a disc. A transparent cover encloses the top of the fuse body and is visible through another cover screwed on to the fuse holder.

The fuse is secured in the fuse holder by a bayonet type fixing, and electrical connection to the charge is by way of terminals in the fuse holder, contact at the base of the fuse and the metal disc.

When current flows in an extinguisher cartridge circuit, the appropriate fuse charge is fired, thereby displacing the disc and interrupting the circuit. At the same time red powder is spattered on to the inside of the cover thus giving a positive visible indication of the firing of the extinguisher cartridge.

Ice and Rain Protection Systems

Icing on aircraft is caused primarily by the presence in the atmosphere of supercooled water droplets, i.e. droplets at a temperature below that at which water normally freezes. In order to freeze, water must lose heat to its surroundings, thus when it strikes, say, an aircraft wing, an engine air intake or a propeller, there is metal to conduct away the latent heat and the water freezes instantly. The subsequent build-up of ice can change the aerodynamic shape of the particular form causing such hazardous situations as decrease of lift, changes of trim due to weight changes, loss of engine power and damage to turbine engine blading. In addition, loss of forward vision can occur due to ice forming on windshield panels, and on externally mounted units such as pitot probes, obstruction of the pressure holes will result in false readings of airspeed and altitude. Therefore, for aircraft which are intended for flight in ice-forming conditions, protective systems must be incorporated to ensure their safety and that of the occupants. Figure 10.20 indicates the extent to which protection may be required depending on the type and size of aircraft.

In addition to ice protection systems, some

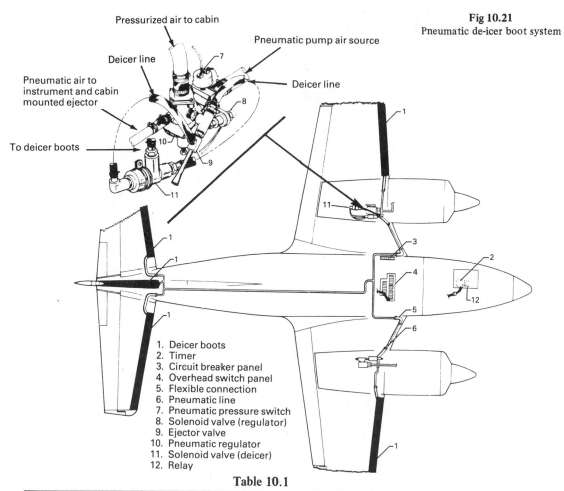

Fig 10.21
Pneumatic de-icer boot system

Pressurized air to cabin

Pneumatic pump air source

Deicer line

Pneumatic air to instrument and cabin mounted ejector

Deicer line

To deicer boots

1. Deicer boots
2. Timer
3. Circuit breaker panel
4. Overhead switch panel
5. Flexible connection
6. Pneumatic line
7. Pneumatic pressure switch
8. Solenoid valve (regulator)
9. Ejector valve
10. Pneumatic regulator
11. Solenoid valve (deicer)
12. Relay

Table 10.1

Method	Application	Principle
FLUID	Wings, stabilizers, propellers, windshields	A chemical which breaks down the bond between ice and water and can be either sprayed over the surface, e.g. a windshield, or pumped through porous panels along the leading edge of a surface, e.g. a wing.
PNEUMATIC BOOT	Wings, stabilizers	Sections of rubber boot along the leading edges are inflated and deflated causing ice to break up and, with aid of the airstream, crack off.
THERMAL (a) Hot air bleed	Wings, stabilizers, engine air intakes	Hot air from turbojet engine compressors passed along inside of leading edge structure.
(b) Combustion heating	Wings, stabilizers	Hot air from a separate combustion heater or from a heat exchanger associated with a turbine engine exhaust gas system.
(c) Electrical heating	Wings, stabilizers, engine air intakes, propellers, helicopter rotor blades, windshields	Heating effect of electric current passing through wire, flat strip or film type elements.

De-icing is effected by de-icer "boots" which are cemented to the leading edges of the appropriate surfaces, and which, at controlled time intervals, inflate to break up ice which has formed on them. In some systems the boots are inflated in a specific sequence, but in the example shown, all the boots inflate simultaneously for a period of 6 seconds. The boots are of fabric-reinforced rubber and contain built-in inflation tubes arranged spanwise (see Fig. 10.22) which are connected to an air supply system via solenoid-operated valves. In some types of aircraft, the tubes are arranged chordwise so that they minimize interference with the airflow over the relevant surfaces. A thin conductive coating is provided over the surfaces of the boots to dissipate static charges.

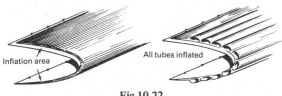

Fig 10.22
De-icer boot operation

The air supply in the case of the aircraft considered is derived from engine-driven pumps and is regulated at 18 psi, but in aircraft powered by turbo-propeller engines, the air is usually tapped from an engine main compressor stage and then regulated to the desired pressure. During the deflation period of the operating cycle, and also during flight under no-icing conditions, it is necessary for the boots to be held flat against the leading edges of the appropriate surfaces, and this is achieved by connecting the boots, via their solenoid-operated valves, to a vacuum source. This is derived by passing air continuously overboard from the engine-driven pumps or, engine compressor as the case may be, through an ejector/venturi.

ELECTRICAL DE-ICING AND ANTI-ICING SYSTEMS

It is beyond the scope of this book to go into the construction and operating detail of any one specific system, but the following details, although of a general nature only, may nevertheless, be considered as typical.

A system is made up of three principal sections: heating elements, control, protection and indicating. The power supplies normally required are 115 volts to 200 volts a.c. for heating (although the propellers for some light aircraft types and some windshield panels operate on 28 volts d.c.), 115 volts a.c. and 28 volts d.c. for control and for other sections of a system. Depending on the application, heating current may be controlled to permit de-icing, anti-icing or both.

The heating elements vary in design and construction depending on the application. For propellers they are of the fine wire type sandwiched in insulating and protective materials which form overshoes selected for maximum resistance to environmental conditions and bonded to the blade leading edges. For propeller-turbine engine air intakes, leading edges of wings, and helicopter rotor blades, the elements are of the "sheared foil" type, i.e. they are cut from thin sheets of high-grade metal to specified lengths and widths and within very close tolerances. The final resistance values of the elements which are selected from such metals as nickel, copper-nickel and nickel-chrome, are usually adjusted by chemical etching. The elements are also sandwiched between insulating and protective layers to form overshoes or mats.

Figure 10.23 illustrates one example of a propeller and air intake de-icing system. Electrical power, at 200 volts a.c and variable frequency, is supplied to the propeller blades and spinner, via brushes and slip rings and a cyclic time switch, so that during the

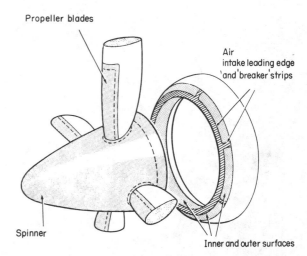

Fig 10.23
Propeller and air intake de-icing system

de-icing part of the cycle, heat is applied to all four blades simultaneously. It is unnecessary to de-ice the whole of each blade, as kinetic heating allied to centrifugal force normally keeps the outer halves free from ice.

The air intake elements are arranged so that those positioned at the leading edges are continuously heated, i.e. they perform an anti-icing function, while those on the inner and outer surfaces are supplied via the cyclic time switch and so perform a de-icing function. In order that ice may be shed in reasonably-sized sections, the leading edge heating elements are extended at intervals to form "breaker" strips. The resistance of the elements is graded to provide for various heating intensities required at different parts of the air intake.

The heating element arrangements adopted in another current type of turbopropeller engine, are shown in Figs. 10.24 and 10.25. In this case the power supply for heating is 28 volts d.c.

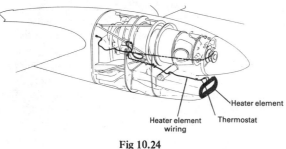

Fig 10.24
Engine air intake anti-icing

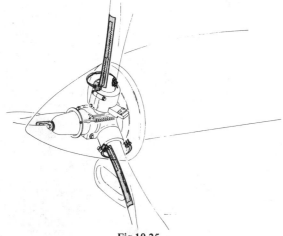

Fig 10.25
Electrical heater elements – propeller de-icing

For windshields or other essential clear vision panels in cockpits, a transparent metal film type of element is employed in the majority of applications, the metal being either stannic oxide or gold. Panels are of laminated construction, and in order to provide rapid heat transfer the metal film is electrically deposited on the inside of the outer glass layer. It is protected from damage and completely insulated by further layers of polyvinyl butyral, glass and/or acrylic. Heating current, normally from an a.c. source, is supplied to the film by metal busbars at opposite edges of the glass layer. The power necessary to deal with the most severe icing conditions is in the order of 5–6 watts/in^2 of windshield area.

In systems applied to the windshields of some small types of aircraft, heating elements made up of fine resistance wire are used and are connected to a 28-volts d.c. power source.

Windshield systems are essentially anti-icing systems for, in addition to the protective function, the temperature of the panels must be higher than ambient during take-off, flight at low altitudes and landing, thus making them "pliable" and thereby improving their impact strength against possible collision with birds.

Temperature Control Methods

In view of the high amounts of power required for the foregoing electrical heating methods, it is essential to provide each system with appropriate controlling circuits and devices. Although there are a number of variations between systems and between designs adopted by different manufacturers, from the point of view of primary functions they are more or less the same, i.e. to cycle the power automatically, to detect any overloading and to isolate power supplies under specific conditions.

ENGINE AIR INTAKE AND PROPELLER SYSTEMS

Figure 10.26 represents the supply and control circuit for the engine air intake and propeller shown in Fig. 10.23.

When the system is switched on, direct current energizes the power relay via closed contacts in the overload sensing device, thus allowing the 200 volts a.c. to flow directly through to the continuously heated elements and up to the time switch (see also p. 104). This unit is energized to run either "fast" or "slow" by a selector switch, the settings being governed by outside air temperature and severity of icing. In this case, "fast" is selected at temperatures between +10°C and −6°C and the duration of the "heat on" and "heat off" periods of the cyclic heated elements is short compared with "slow", which is selected at temperatures below −6°C. The cycling is usually controlled by cam-operated microswitches. An indication of time switch operation is provided by a flashing blue or green light on the control panel, while a general indication that the correct power is being applied to the whole system is provided by an ammeter connected to a current transformer (see also p. 61) across the generator busbar.

In the event of an a.c. overload, the heater elements are protected by the sensing device which is actuated in such a manner that it interrupts the d.c. supply to the power relay, this in turn interrupting the supply of heating current. The current balance relay fulfils a similar function and is actuated whenever there is an unbalance between phases beyond a predetermined amount.

For ground operation of the system described, it is usual for the applied voltage to be reduced in order to prevent overheating. This is effected by the automatic closing of a microswitch fitted to a landing

gear shock-strut, the switch permitting direct current to flow to a reduced voltage control section within the generator voltage regulator.

The circuit of a d.c. powered system as applied to the air intake illustrated in Fig. 10.24, is shown in Fig. 10.27. When the system is switched on, the control relay is energized by the 28-volt supply having to pass to ground through the closed contacts of a thermostat and through the contacts of the engine oil pressure switch which closes when the pressure reaches 50 psi. The supply for the heater elements which is rated at 500 watts, then passes through the energized relay contacts.

The thermostat prevents overheating of the heater element by opening the circuit when the element temperature reaches $49 \pm 3\,°C$ $(120 \pm 5\,°F)$. The oil pressure switch opens the circuit when as a result of engine shutdown the oil pressure falls below 50 ± 2 psi. Functioning of the heating circuit is indicated by an annunciator light on the system control panel. The light is illuminated by a current-sensing relay which is in series with the heater element, and energizes when heater current is above 15 amperes.

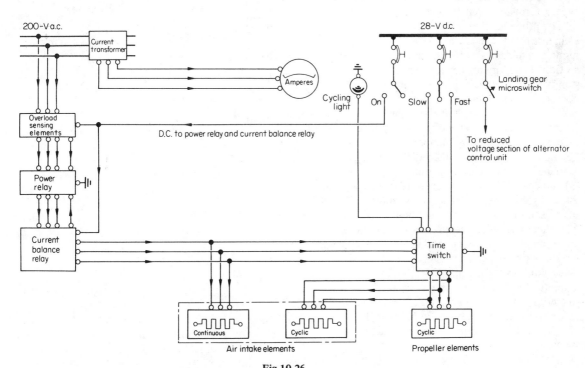

Fig 10.26
Engine air intake and propeller de-icing and anti-icing
control circuit

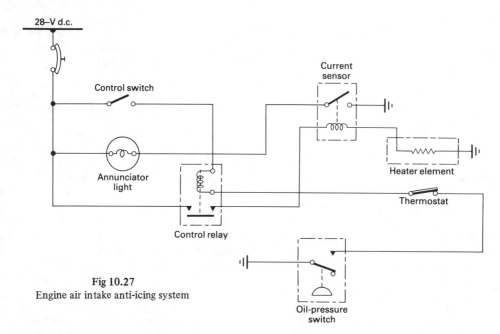

Fig 10.27
Engine air intake anti-icing system

An example of a propeller de-icing system utilizing 28-volt d.c. power for heating and control is shown in Fig. 10.28, and is one which is applied to several types of small twin-engined aircraft.

The propeller blades each have two heater elements bonded to them; one at the outboard section of a blade and the other at the inboard section. The elements are connected to the power supply via slip rings, brushes and an electrically-operated timer which is common to both propellers.

The cycling sequence of the timer is set so that – (i) the outboard elements of each propeller are simultaneously heated before the inboard elements, and (ii) only one propeller is de-iced at a time. The sequence for the right-hand propeller is shown at (a) and (b) of Fig. 10.28 respectively. The segments 3 and 4 respectively connect the supply to the outboard and inboard elements of the left-hand propeller. The timer energizes the elements for approximately 34 seconds and repeats the cycle as long as the control switch is in the "on" position. Operation of the system is indicated by the ammeter, the pointer of which registers within a shaded portion of the ammeter scale corresponding to current consumed (typically between 14 and 18 amperes) at the normal system voltage.

WINDSHIELD ANTI-ICING SYSTEMS
The control methods adopted for windshield anti-icing systems are normally thermostatic, and a typical

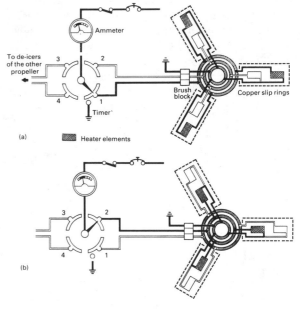

Fig 10.28
Propeller de-icing system

system (Fig. 10.29) consists of a temperature-sensing element and a control unit. The element is embedded within the panel in such a way that it is electrically insulated from the main heating film and yet is capable of responding to its temperature changes without any serious lag. A control unit comprises mainly a bridge circuit, of which the sensing element forms part, an amplifier and a relay. When all the required power is switched on initially, the control unit relay is energized by an unbalanced bridge signal and the power control relay is energized to supply the windshield panel. As the panel temperature begins to increase, the sensing element resistance also increases until at a predetermined controlling temperature (a typical value is 40°C) the current flowing through the sensing element balances the bridge circuit, and the control unit and power control relays are de-energized, thereby interrupting the heating current supply. As the temperature cools the sensing element resistance decreases so as to unbalance the bridge circuit and thereby restore the heating current supply. In a number of aircraft types the windshields are each fitted with an additional overheat sensing element which in the event of failure of the normal sensing element takes over its function and controls at a suitably higher temperature; 55°C is a typical value.

Despite accurate control during manufacture slight variations in heater film resistance, and consequently glass temperature, can occur. Sensing elements are,

therefore, individually embedded in each panel at one of the hotter spots but where it least affects visibility.

In some types of aircraft, windshields are heated by resistance elements of fine wire supplied with 28 volts d.c. Temperature sensing and control of heating current is carried out by a control unit operating on a similar principle to that already described.

Hot-Air Bleed Anti-Icing Systems

Systems of this type are standard principally on the larger types of public transport aircraft, for the anti-icing of engine air intake nose cowlings, wing leading edges and leading edge devices such as slats and flaps (see Appendix 9).

The hot air is bled from certain stages of main engine compressors and is then ducted through metal ducting to the air intakes and leading edges. As far as the use of electrical power is concerned, this is required solely for the operation of motorized control valves in the ducting, valve position indicating lights, and duct temperature sensing devices. The motors are limit switch controlled at the full open and closed positions, and in most applications they

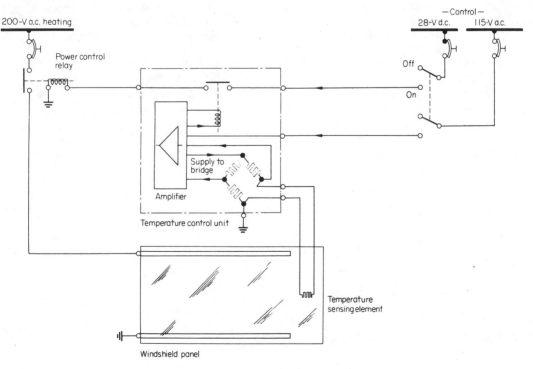

Fig 10.29

Schematic arrangement of windshield anti-icing control system

174

are of the 115-volts single-phase a.c. type. A power supply of 28 volts d.c. is used for valve control relay switching and position indicating light circuits.

The d.c. supply for valve control relay switching passes through a landing gear shock-strut micro-switch so that when an aircraft is on the ground the anti-icing system cannot be operated as in the normal in-flight situation. A ground test switch circuit is therefore provided to check the operation of valves and position indicating lights.

Ice Detection Systems

These systems consist mainly of a sensing probe located at a strategic point on an aircraft (usually the front fuselage section) and a warning light, their purpose being to give adequate warning, and an assessment of the likely severity of an impending icing hazard in sufficient time for the ice protection systems to be brought into operation. Detectors are made in a variety of forms, and in those most commonly used actuation of the warning circuit is triggered off by ice accretion at the sensing probe.

In one type of system ice accretion causes a drop in pressure sensed by the probe and a diaphragm, the deflections of which make a circuit to the warning light and to a heater within the probe. When the ice has melted the warning light and heater circuits are interrupted and the system is reset for further ice detection.

A second type of system is designed to give a warning and also automatically switch on airframe and engine de-icing systems. It consists of an a.c. motor-driven rotor which rotates in close proximity to a knife-edge cutter, a time delay unit and a warn-ing lamp. Under icing conditions ice builds up on the rotor and closes the gap between it and the cutter. This results in a substantial increase in the torque-loading on the detector motor, causing it to rotate slightly in its mounting and to trip a microswitch inside the detector. Tripping of the microswitch completes the circuit to the warning light and time delay unit which initiates operation of the de-icing systems. These conditions are maintained until the icing diminishes to the point whereby the knife-edge cutter ceases to "shave" ice, and the microswitch is returned to the open circuit condition. The detector unit is designed to provide a two minute interval be-tween the cessation of an ice warning and shut down of a de-icing system, to prevent continuous interruption of the system during intermittent icing conditions.

In a third type of system, the fundamentals of operation are dependent on the phenomenon of magnetostriction, i.e. its sensing probe is caused to vibrate axially when subjected to a magnetic field at specific frequencies. The function of the system is shown in Fig. 10.30.

The sensing probe is a ¼-inch diameter nickel alloy (Ni-Span C) tube mounted at its mechanical centre. The inherent resonant frequency of the probe is inversely proportional to its length, the simplified relationship being expressed as $f = \dfrac{S}{2L}$ where,

f = frequency in Hz, $S = 1.88 \times 10^5$ inches per second (the speed of sound in the tube material) and L = length of the probe in inches. Based on this expression, the tube may be cut to a specific length to achieve a desired frequency; in this particular system the designed tube length is 2·3 inches, resulting in a

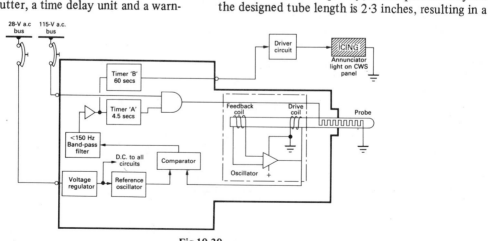

Fig 10.30
Functional diagram of ultrasonic probe system

resonant ultrasonic frequency of 41 kHz. This frequency, however, is reduced to a nominal 40 kHz by the brazing of heating elements within the tube and also by capping the tip of the tube. The probe is maintained in its axial vibration by the ultrasonic frequency excitation current produced by an oscillator and passed through a drive coil wound around the probe. The frequency is controlled by a feedback coil circuit such that the drive coil will excite the probe at whatever the natural frequency of the probe might be at the time. When ice forms on the probe the natural frequency is reduced, and the output frequency of the oscillator drive coil is in turn reduced to match the probe frequency. By means of a comparator circuit, the lower output frequency is compared with a fixed frequency output from a reference oscillator. The frequency difference between the two oscillators relates to the ice formation on the probe, and when the difference has reached a preset level (150 Hz or less) determined by a band pass filter and a limiting amplifier, a signal is sent to a switch and delay circuit. When this occurs, two timer circuits are triggered; one controlling the a.c. supply via a logic AND gate, to the probe heater, and the other controlling the duration an icing signal is available to an annunciator light for warning the flight crew. Thus, as will be noted from Fig. 10.30, there is a standing logic 1 input to the AND gate from the 115-volt bus, so when timer "A" is triggered it will supply a second logic 1 input to the gate causing it to switch on the heater for a period of 4.5 seconds. The signal from timer "B" is 28 volt d.c. and keeps the annunciator light illuminated for a period of 60 seconds. Melting of the ice from the probe increases the frequencies of the probe, and if no other icing signal is detected within 60 seconds, timer "A" automatically resets to isolate the heater from the a.c. supply. This cycle of operation is repeated while icing conditions prevail.

Failure monitoring of the detector is accomplished with unijunction oscillators which are set at both ends of the maximum difference frequency band. If the probe becomes severely damaged causing a significant change in the resonant frequency, or if an electronic component failure causes a malfunction in the reference frequency circuit, the annunciator light will be continuously illuminated.

Landing Gear Control

In a number of the smaller types of aircraft having a retractable landing gear system, the extension and

retraction of the main wheels and nose wheel, is accomplished by means of electrical power. Fig. 10.31 is a simplified circuit diagram of a representative control system.

The motor is of the series-wound split-field type (see also pp. 137 and 138) which is mechanically coupled to the three "leg" units, usually by a gearbox, torque shafts, cables, and screw jacks. The 28 volts d.c. supply to the motor is controlled by a selector switch, relay, and switches in the "down-lock" and "up-lock" circuits. A safety switch is also included in the circuit to prevent accidental retraction of the gear while the aircraft is on the ground. The switch is fitted to the shock-strut of one of the main wheel gear units, such that the compression of the strut keeps the switch contacts in the open position as shown in the diagram.

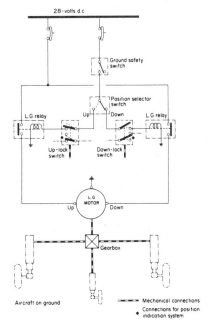

Fig 10.31
Landing gear control system

After take-off, the weight of the aircraft comes off the landing gear shock-struts, and because they have a limited amount of telescopic movement, the strut controlling the safety switch causes it to close the switch contacts. Thus, when the pilot selects "gear up", a circuit is completed via the selector switch, and closed contacts of the up-lock switch, to the coil of the relay which then completes the supply circuit to the "up" winding of the motor. When the landing gear units commence retracting, the down-lock switch is automatically actuated such that its contacts will also close, and will remain so up to and in the fully retracted position. As soon as this position is reached, the up-lock switch is also actuated so as to open its contacts, thereby interrupting the supply to the motor, and the "down" winding circuit of the motor is held in readiness for extending the landing gear. As and when the appropriate selection is made, and the landing gear units commence extending, the up-lock switch contacts now close and when the landing gear is down and locked, and the aircraft has landed, the circuit is again restored to the condition shown in Fig. 10.31.

To prevent over-run of the motor, and hence over-travel of the landing gear units, some form of braking is necessary. This is accomplished in some cases, by incorporating a dynamic brake relay in the circuit. The relay operates in such a manner that during over-run, the motor is caused to function as a generator, the resulting electrical load on the armature stopping the motor and gear instantly.

Landing Gear Position Indication

In retractable landing gear systems, it is, of course, necessary to provide some indication that the main

and nose landing gear units are locked in their retracted positions during flight, and in their extended positions safe for landing. The indication method most widely adopted is based on a system of indicating lights which are connected to microswitches actuated by the up-lock and down-lock mechanisms of each landing gear unit. To guard against landing with the landing gear retracted or unlocked, a warning horn is also incorporated in the indication system. The horn circuit is activated by a microswitch the contacts of which are made or broken by the engine throttle. Fig. 10.32 illustrates a typical circuit arrangement.

The system operates from a 28 volts d.c. power supply which is connected to lamps within the indicator case, and also to the up-lock and down-lock microswitches of the main and nose landing gear units. Three of the lamps are positioned behind red screens, and three behind green screeens; thus, when illuminated they indicate respectively, "gear up and locked" and "gear down and locked". In the "gear up and locked" position all lights are extinguished. In the event of failure of a green lamp filament, provision is made for switching-in a standby set of lamps.

The circuit as drawn, represents the conditions when the aircraft is on the ground in a completely static condition. As soon as power goes onto the busbar, the three green lamps will illuminate because

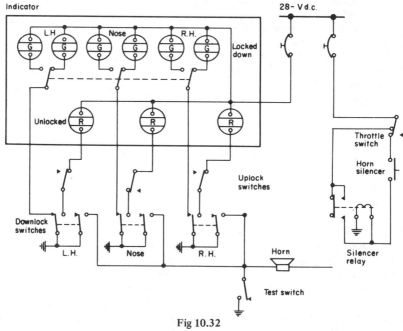

Fig 10.32
Landing gear position indicating system

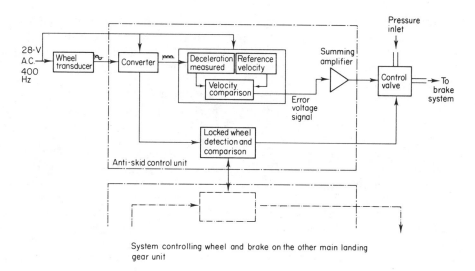

Fig 10.33
Anti-skid control system

their circuits are completed to ground via the left-hand set of contacts of the corresponding down-lock microswitches. The engine throttle is closed, and although its microswitch is also closed, the warning horn circuit is isolated since there is no path to ground for current from the busbar. Assume now that the aircraft has taken off and the pilot has selected "landing gear up"; the down-lock mechanisms of the gear units are disengaged and they cause their microswitches to change contact positions, thus interrupting the circuits to the green lamps. At the same time, the red lamps are illuminated to indicate that the gear units are unlocked, the power supply for the circuit passing to ground via the up-lock switches, and the right-hand contacts of the down-lock switches. When the landing gear units reach their retracted positions, the up-lock mechanisms are engaged and cause their microswitches to interrupt the circuits to the red lamps; thus, all lamps are extinguished. When the pilot selects "landing gear down", the up-lock mechanisms now disengage and the microswitches again complete the circuit to the red lamps to indicate an unlocked condition. As soon as the gear units reach the fully extended position, the down-lock mechanisms engage and their microswitches revert to the original position shown in Fig. 10.32 i.e., red lamps extinguished, and green lamps illuminated to indicate "down and locked".

As noted earlier, a warning horn is included in the system, the making and breaking of the horn circuit being controlled by a throttle-operated microswitch.

In the static condition shown in Fig. 10.32, the throttle microswitch is closed, but the warning horn will not sound since the circuit is interrupted by all three down-lock microswitches. Similarly, the circuit will be interrupted by the throttle microswitch which is opened when the throttle is set for take-off and normal cruise power. In the case of an approach to land, the engine power is reduced by closing the throttle to a particular approach power setting and this action closes the throttle microswitch. If, in this flight condition, the landing gear has not been selected down in readiness for landing, then the warning horn will sound since the circuit to ground is then completed via the right-hand contacts of the down-lock microswitches. After selecting "down", the horn continues to sound, but it may be silenced by operating a push switch which, as will be noted from the diagram, energizes a relay to interrupt the horn circuit. The relay incorporates a hold-in circuit so that it will remain energized until the d.c. power supply is finally switched off. Functional testing of the horn circuit on the ground, and under engine static conditions, may be carried out by closing the throttle and its microswitch, and then operating a test switch.

Anti-Skid Control Systems

The braking systems of many large transport aircraft are provided with a means of preventing the main landing gear wheels from skidding on wet or icy surfaces, and of ensuring that optimum braking effect can be obtained under all conditions, by modulating the hydraulic pressure applied to the brakes. Fundamentally, an anti-skid system senses the rate of change of wheel deceleration, decreases the hydraulic pressure applied to the brakes when there is an impending skid condition, and restores the pressure as the wheel accelerates again.

A number of current anti-skid systems utilize electrical power, and typically a complete system consists of a number of transducers (one for each wheel of each main landing gear unit), an anti-skid control unit containing the requisite number of individual circuits, and electro-hydraulic anti-skid control valves corresponding to the number of transducers, and control circuits. A block diagram of one such system is illustrated in Fig. 10.33; for simplicity of explanation the diagram and description that follows, relate to single wheel operation.

The transducer is a speed sensing device and consists of a stator which is firmly attached to the wheel axle, and a rotor which is attached to, and rotates with, the wheel. The stator contains a permanent magnet, and when the wheel and rotor are rotated, the magnetic coupling, or magnetic reluctance, between the rotor and stator, is varied. The variation generates within the stator, an a.c. voltage signal which is directly proportional to the rotational speed of the wheel. The signal is fed to the converter in the control unit, and is converted to a d.c. voltage signal which serves as a measure of the rate of wheel deceleration. The signal is then applied to a skid control circuit and is compared with a reference velocity signal which has been predetermined from a known deceleration rate of the aircraft. Any differences between the two signals produce error voltages which are processed to determine whether or not a correction signal is to be applied to the electro-hydraulic control valve. If wheel deceleration rates are below the reference velocity no correction signal is produced. If, however, the rates are above the reference velocity, they are then treated as skids or approaching skids, and correction signals are applied to the control valve which reduces the hydraulic pressure applied to the wheel brakes. When the wheel speed falls below the reference deceleration rate, the skid control unit transmits a release signal to the control valve. Subsequent wheel "spin-up" causes the brakes to be re-applied, but at a lower pressure determined by the length of time required for the wheel to spin-up. Sensing circuits are also provided, and in conjunction with the systems of the other wheels of the landing gear, they detect and compare "locked wheel"

conditions. In the event of such conditions occurring, the circuits will cause signals to be applied to the relevant control valves such that they will fully release brake pressure.

Windshield Wiper Systems

The circuit arrangement shown in Fig. 10.34 is typical of many of the windshield wiper systems currently in use. The wiper arms and blades for each windshield are actuated by their own 28-volt d.c. variable-speed motors coupled to converters. Each motor is supplied from different busbars and is controlled by a four-position selector switch (in some cases the switch may have six positions) and the speed variation according to selection, is accomplished by voltage dividing resistances.

In the "low" position of the switch, voltage is applied to the field and armature circuits of the motor, and then to ground via a second contact of the switch and two resistors. The voltage is therefore reduced and the motor runs at a low speed, and by means of its converter sweeps the wiper arm back and forth. When the "high" position of the switch is selected, the supply passes to ground through only one resistor and so the motor and wiper will operate at a faster speed.

When the use of the wipers is no longer required, the control switch is turned back through the "off" position to a "park" position. There is no detent in this position, and so the switch is manually held there momentarily. As will be noted from Fig. 10.34, the supply voltage is initially applied to the motor in the normal way, but as the connection to ground is now directly through the normally-closed contacts of a brake switch within the motor, then it will run at its fastest speed. As the wiper blade reaches its parked position, the motor operates a cam to change over the brake switch contacts which then short out the armature to stop the motor. The switch is then released to spring back to the "off" position.

The purpose of the thermal switch is to open the motor circuit if the field winding temperature or field current should exceed pre-determined values. Typical values are 150°C (300°F) and 8 to 10 amperes respectively.

Rain Repellent Systems

The purpose of these systems is to maintain a clear area on the windshields of an aircraft during take-off, approach and landing in rain conditions. A system consists of a pressurized container of repellent

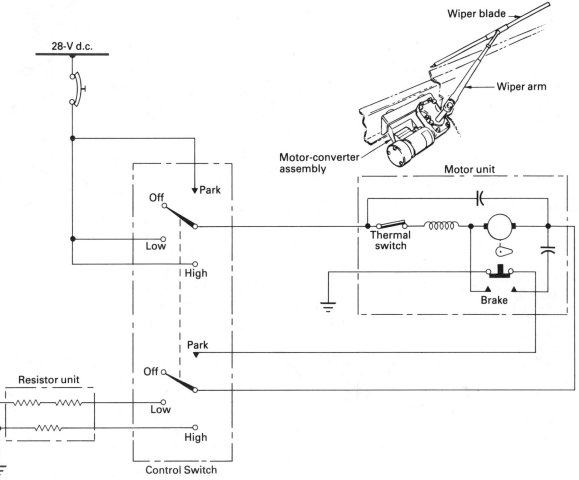

Fig 10.34
Windshield wiper system

fluid, control switch, a solenoid valve controlling the supply of fluid to a spray nozzle mounted in the fuselage skin in front of each windshield. The fluid container is common to each windshield system and is located in the cockpit.

The operation of the system is illustrated in Fig. 10.35. When the control switch is pushed in, a 28-volt d.c. supply is fed to the solenoid valve via the closed contacts "B" of the control relay. The spray nozzle solenoid is therefore energized to open the valve and allow fluid to flow under pressure through the spray nozzle and onto the windshield. The fluid is of a type which causes the surface tension in water to change so that the water is formed into globules which are blown off the windshield by the airstream. Through the action of a time delay circuit,

approximately 5 c.c. of fluid flows through the nozzle for approximately 0.25 seconds.

At the end of this period, the time delay circuit applies power to the gate of an SCR (see also p. 55) which then energizes the control relay and in turn de-energizes the spray nozzle solenoid valve. If the control switch remains pushed in, the time delay circuit will keep the control relay energized via a hold-in circuit across the closed contacts "A". When the switch is released, the time delay circuit and SCR are returned to their original state.

The fluid is contained in a can which when screwed onto the mounting bracket opens a valve to allow fluid to drain into a reservoir and the system tubing. The reservoir is a clear plastic cylinder containing a float-type contents indicator. A manually-

180

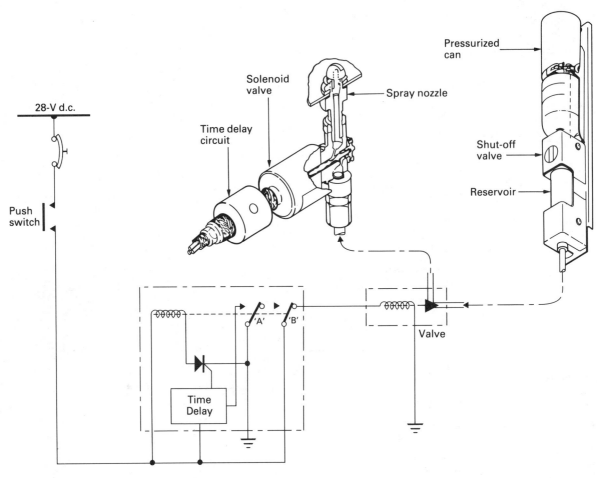

Fig 10.35
Rain repellent system

operated shut-off valve is provided between the
reservoir and can and is used during can replacement.

Airconditioning Systems

These systems are designed to maintain selected air
temperature conditions within flight crew, passenger
and other compartments, and in general, they are
comprised of five principal sections: air supply,
heating, cooling, temperature control, and distribution. The operation of systems varies depending
on the size and type of aircraft for which they are
designed, and space does not allow for them all to
be described. However, if we take the case of most
of the large transport aircraft, we find that there
are a number of common features which may be
represented as shown in Fig. 10.36.

As in the case of hot air bleed anti-icing systems,
air is supplied from stages of the main engine compressors and serves not only to provide air conditioning but also pressurization of the cabin. Since
the air from the compressor stages is too warm for
direct admission to the cabin, it has to be mixed
with some cold air in order to attain preselected
temperature conditions. This is effected by directing
some of the bleed air through a cooling pack consisting of a heat exchanger system and a cooling
turbine or air cycle machine. The control of the
bleed air flow is accomplished by an electrically
controlled pack valve, which is energized by a switch
on the system control panel in the cockpit. Down-
stream of the pack valve is a mix valve which has
the function of proportionately dividing the hot

air flow from the pack valve, and the cold air flow from the air cycle machine, into a mixing chamber. The mix valve is of the dual type; both valves being positioned by a common 115-volt a.c. actuator motor. The valves are monitored by signals from the temperature control system such that as one valve moves towards its close position, the other valve moves towards its open position.

The temperature control system is comprised principally of a selector switch, regulator, and temperature sensors located at selected points in the system. The whole system operates automatically and continuously monitors the mix valve position, but in the event of failure of the regulator, mix valve position may be carried out manually from the selector switch.

When the selector switch is in the "auto" position and at a desired cabin temperature, a potentiometer within the switch establishes a reference resistance value in an arm of a control bridge circuit of the regulator. A cabin temperature sensor is in the other arm of the control bridge circuit so that if the

cabin temperature is at a level other than that selected, then the bridge will be unbalanced. As a result, a signal is developed in the circuit of the mix valve motor so that it will drive the valves to either a hot or cold position, as required, to attain the selected cabin temperature. At the same time, conditioned air is sensed by an anticipator sensor, and a limit sensor both of which are located in the ducting to the cabin, and are connected in an electrical bridge configuration. The purpose of the sensors is to modulate any rapid changes demanded by an unbalanced control bridge so that when the actuator control moves the mix valve it will produce cabin temperature changes without sudden blasts of hot or cold air, and without raising duct temperatures above limits.

To prevent the mix valve staying at a "too hot" position, a thermal switch which is set at a particular level (e.g. 90 °C (195 °F)) is located in the ducting to complete a circuit to the mix valve so that its motor will run the valve to the full cold position. At the same time a "duct overheat" light is

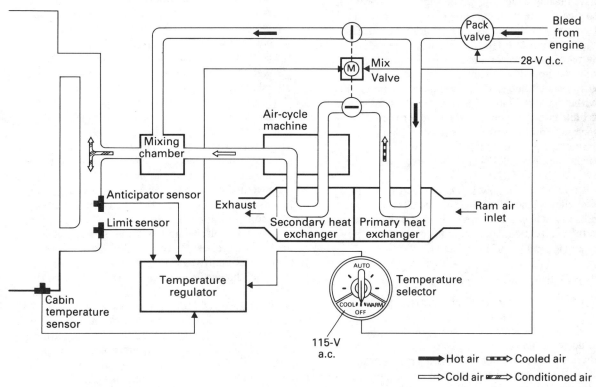

Fig 10.36
Airconditioning system

illuminated. After the overheat condition has been corrected, the system may be returned to normal by means of a reset switch. Another thermal switch set to close at a higher level (e.g. 120 °C (250 °F)) protects against duct overheat should power control be lost. It completes a circuit which closes the pack valve and illuminates a "pack trip off" light. The system may be returned to normal after the trip condition has been corrected, by operating the reset switch referred to above.

Manual control of the system is effected by moving the selector switch to "cool" or "warm" to directly actuate the mix valve as appropriate.

Propeller Synchronizer Systems

These systems are used in some types of twin-engined aircraft, their purpose being to automatically synchronize the r.p.m. of the propellers. This is accomplished by utilizing the speed governor of one propeller as a master unit, and the governor of the second propeller as the slave unit. Both governors have magnetic pick-ups which supply electrical pulses to a control unit which detects any difference in the frequency of the pulses. The resulting output from the control unit is fed to a stepping type motor actuator mounted on the slave governor which is then "trimmed" to maintain its propeller r.p.m. at the same value as the master governor unit, and within a limited range. The limiting range of operation is built into the synchronizer system to prevent the slave governor unit from losing more than a fixed amount of propeller r.p.m. in the event of the master engine and propeller being "feathered" when the system is in operation.

Before the system is activated, the r.p.m. of each propeller is manually synchronized as close as possible. When this has been done and the system is then activated, a maximum synchronizing r.p.m. range (typically ± 67) is effective.

Passenger Cabin Services

In passenger transport aircraft electrical power is required within the main cabin compartments for the service and convenience of the passengers, the extent of power utilization being governed of course, by the aircraft size and number of passengers it is designed to carry. Apart from the main cabin lighting referred to on page 152 it is necessary to provide such additional services as individual reading lights at each seat position, a cabin attendant call system, public address system and a galley for the preparation and serving

of anything from light refreshments to several full-course meals. In-flight cinema entertainment also accounts for the utilization of electrical power in many types of aircraft.

Reading lights may be of the incandescent or fluorescent type, and are located on passenger service panels on the underside of hat racks, or in each seat headrest and are controlled individually. Cabin attendant call systems are interlinked systems comprising switches at each passenger service panel connected to an electrical chime and indicator light at the cabin attendant's panel station. The service panel switches are of the illuminating type to visually indicate to the cabin attendant the seat location from which a call has been made. In addition the system provides an interconnection between the flight crew compartment and cabin attendant's station.

A public address system is provided for giving passengers instructions and route information, and usually comprises a central amplifier unit and a number of loudspeakers concealed at various points throughout the cabin, and in toilet compartments. Information is given, as appropriate, by the aircraft's captain or cabin attendant by means of separate telephone type handsets connected to the loudspeakers. Tape-recorded music may also be relayed through the system during passenger embarkation and disembarkation.

Galley equipment has a considerable technical influence on the design of an aircraft's electrical system, in that it represents a very high percentage of the total system power requirement, and once installed it usually becomes a hard-worked section of an aircraft. The type of equipment and power loadings are governed by such factors as route distances to be flown, number of passengers to be carried and the class configurations, i.e. "economy", "first-class" or "mixed". For aircraft in the "jumbo" and "wide-bodied" categories, galley requirements are, as may be imagined, fairly extensive. In the Boeing 747 for example, three galley complexes are installed in the cabin utilizing both 28 volts d.c. and 115 volts a.c. power and having a total power output of 140 kVA; thus, assuming that the generator output is rated with a power factor of unity, the equivalent d.c. output is 140 kilowatts or in terms of horsepower approximately 187! The galley unit of the wide-bodied Lockheed "Tristar" is also a complex unit but is located as a central underfloor unit. It also utilizes d.c. and a.c. power not only for heating purposes but also for the operation of lifts which transport service trolleys to cabin floor level.

The equipment varies, some typical units being containers and hot cups for heating of beverages, hot cupboards for the heating of pre-cooked meals and ovens for heating of cold pre-cooked meals, a number of which may have to be served, e.g. on long-distance flights. Other appliances required are water heaters for galley washing-up and toilet washbasins, and re-frigerators. In most cases, the equipment is assembled as a self-contained galley unit which can be "plugged in" at the desired location within the aircraft.

It is usual for the electrical power to be supplied from the main distribution systems, via a subsidiary busbar and protection system, and also for certain galley equipment to be off-loaded in the event of failure of a generating system. The load-shedding circuit is automatic in operation and any override system provided is under the pilot's control; on some aircraft load-shedding is also controlled via a landing gear shock-strut microswitch thereby conserving electrical power on the ground. The control panel or panels, which may be mounted on or adjacent to the galley unit, incorporates the control switches, indicator lights and circuit breakers associated with each item of galley equipment, and also the indicator lights of the cabin attendant call system.

CHAPTER ELEVEN

Electrical Diagrams and Identification Schemes

As in all cases involving an assembly, interconnection or maintenance, of a number of components forming a specific system, a diagram is required to provide the practical guide to the system. Aircraft electrical installations are, of course, no exception to this requirement and the relevant drawing practices are specialized subjects necessitating separate standardization of detail to ensure uniformity in preparation and presentation. The standards to which all diagrams are normally drawn are those laid down by appropriate national organizations, e.g. the British Standards Institution, Society of British Aerospace Constructors (S.B.A.C.) and in Specification 100 of the Air Transport Association (A.T.A.) of America. The ATA 100 system has a much greater application internationally than any other. There are usually three types of diagram produced for aircraft namely, circuit diagrams, wiring diagrams and routing charts.

Circuit Diagrams. These are of a theoretical nature and show the internal circuit arrangements of electrical and electronic components both individually and collectively, as a complete distribution or power consumer system, in the detail necessary to understand the operating principle of the components and system. Circuits are normally drawn in the "aircraft-on-the-ground" condition with the main power supply off. In general, switches are drawn in the "off" position, and all components such as relays and contactors are shown in their demagnetized state. Circuit breakers are drawn in the closed condition. In the event that it is necessary to deviate from these standard conditions, a note is added to the diagram to clearly define the conditions selected.

Wiring Diagrams. These are of a more practical nature in that they show how all components and cables of each individual system making up the whole

installation, are to be connected to each other, their locations within the aircraft and groups of figures and letters to indicate how all components can be identified directly on the aircraft.

Routing Charts. These charts have a similar function to wiring diagrams, but are set out in such a manner that components and cables are drawn under "location" headings so that the route of distribution can be readily traced out on the aircraft. In some cases, both functions may be combined in one diagram (see Fig. 11.1).

Wiring diagrams and routing charts are provided for the use of maintenance engineers to assist them in their practical tasks of testing circuits, fault finding and installation procedures. The number of diagrams or charts required for a particular aircraft, obviously depends on the size of the aircraft and its electrical installation, and can vary from a few pages at the end of a maintenance manual for a small light aircraft, to several massive volumes for large transport aircraft.

Coding Schemes

As an aid to the correlation of the details illustrated in any particular diagram with the actual physical conditions, i.e. where items are located, sizes of cables used, etc., aircraft manufacturers also adopt an identification coding scheme apart from those adopted by cable manufacturers. Such a scheme may either be to the manufacturer's own specification, or to one devised as a standard coding scheme. In order to illustrate the principles of schemes generally, some example applications of one of the more widely adopted coding standards will be described.

In this scheme, devised by the Air Transport Association of America under Specification No. 100, the

185

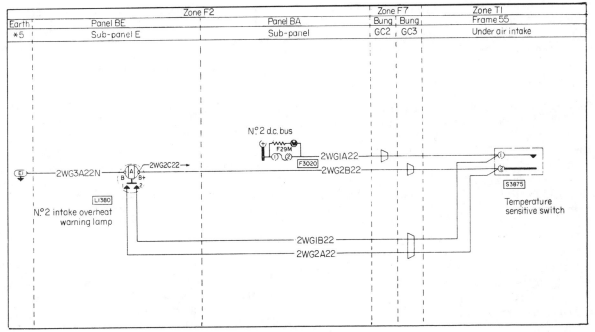

Fig 11.1
Routing chart

coding for cable installations consists of a six-position combination of letters and numbers which is quoted on all relevant wiring diagrams and routing charts and is imprinted on the outer covering of cables. In cases where the code cannot be affixed to a cable it is printed on non-metallic sleeves placed over the ends of the cable. The code is printed at specified intervals along the length of a cable by feeding it through a special printing machine. The following example serves to illustrate the significance of each position of the code:

$$1 \quad P \quad 1 \quad A \quad 22 \quad N$$

Position 1. The number in this position is called the unit number and is only used where components have identical circuits, e.g., the components of a twin generator system. In this case number 1 refers to the cables interconnecting the components of the first system. The number is omitted from cables used singly.

Position 2. In this position, a letter is used to indicate the function of the circuit i.e., it designates the circuit or system with which the cable is connected. Each

system has its own letter. When the circuit is part of radar, radio, special electronic equipment, a second letter is used to further define the circuit.

Position 3. The number in this position is that of the cable and is used to differentiate between cables which do not have a common terminal in the same circuit. In this respect, contacts of switches, relays, etc. are not classified as common terminals. Beginning with the lowest number and progressing in numerical order as far as is practicable, a different number is given to each cable.

Position 4. The letter used in this position, signifies the segment of cable (i.e., that portion of cable between two terminals or connections) and differentiates between segments in a particular circuit when the same cable number is used throughout. When practicable, segments are lettered in alphabetical sequence (excluding the letter "I" and "O") the letter "A" identifying the first segment of each cable, beginning at the power source. A different letter is used for each of the cable segments having a common terminal or connection.

186

Position 5. In this position, the number used indicates the cable size and corresponds to the American Wire Gauge (AWG) range of sizes. This does not apply to coaxial cables for which the number is omitted, or to thermocouple cables for which a dash (−) is used as a substitute.

Position 6. In this position, a letter indicates whether a cable is used as a connection to a neutral or earth point, an a.c. phase cable, or as a thermocouple cable. The letter "N" indicates an earth-connected cable, the letter "V" indicates a supply cable in a single-phase circuit, while in three-phase circuits the cables are identified by the letters "A", "B" and "C". Thermocouple cables are identified by letters which indicate the type of conductor material, thus: AL (Alumel); CH (Chromel); CU (Copper); CN (Constantan).

The practical application of the coding scheme may be understood from Fig. 11.1 which shows the wiring of a very simple temperature sensing switch and warning lamp system.

The system is related to the No. 2 engine air intake, its circuit function is designated by the letters "WG", and it uses cables of wire size 22 throughout. Starting from the power source i.e., from the No. 2 d.c. busbar, the first cable is run from the fuse connection 2, through a pressure bung to terminal 1 of the switch; thus, the code for this cable is 2 WG 1 A 22. Terminal 1 also serves as a common power supply connection to the contact 2 of the press-to-test facility in the warning lamp; therefore, the interconnecting cable which also passes through a pressure bung, is a second segment cable and so its code becomes 2 WG 1 B 22. Terminal 2 of the switch serves as a common connection for the d.c. output from both contact 1 of the press-to-test facility, and the sensing switch contacts, and as the cables are the second pair in the circuit and respectively first and second segments, their code numbers are 2 WG 2 A 22 and 2 WG 2 B 22. The cable shown going away from the B+ terminal of the lamp, is a third segment connecting a supply to a lamp in a centralized warning system and so accordingly carries the code 2 WG 2 C 22. The circuit is completed via cable number 3 and since it connects to earth it carries the full six-position code; thus, 2 WG 2 A 22 N.

The coding schemes adopted for items of electrical equipment, control panels, connector groups, junction boxes, etc, are related to physical locations within the aircraft and for this purpose aircraft are divided into electrical zones. A reference letter and number are allocated to each zone and also to equipment, connectors, panels etc., so that they can be identified within the zones. The reference letters and numbers are given in the appropriate wiring diagrams and are correlated to the diagrammatic representations of all items. In the aircraft itself, references are marked on or near the related items.

Logic Circuits and Diagrams

The operation of the majority of units comprising electrical systems is largely based on the application of solid-state circuit technology i.e. components such as resistors, capacitors and rectifiers that are normally interconnected as separate discrete components, are all "embedded" in micro-size sections of semiconductor material. Apart from the vast reduction in dimensions, this form of integration also makes possible the production of circuit "packs" capable of performing a vast number of individually dedicated functions. Thus, in knowing the operating parameters of a system overall, and the functions constituent units are required to perform, the complete circuit of a system is built up by interconnecting selected functional packs. The packs consist of basic decision-making elements referred to as logic gates, each performing combinational operations on their inputs and so determining the state of their outputs.

As far as the diagrammatic presentation of the foregoing circuits is concerned, greater use is made of a schematic form depicting interconnected blocks and a variety of special logic symbols, each representing a specific circuit network "hidden away" in the semiconductor material. The study of a system's operation is therefore based more on the interpretation of symbols and the logic state of signal functions at the various interconnections of the circuit, rather than tracing through diagrams that depict all internal circuit details in more theoretical form.

LOGIC GATES

Logic gates are of a binary nature i.e. the inputs and the outputs are in one of two states expressed by the digital notation 1 or 0. Other corresponding expressions are also frequently used as follows:

1 − on; true; high (H); closed; engaged
0 − off; false; low (L); open; disengaged

The 1 and 0 state designations are arbitrary. For example, if the states are represented by voltage levels, one may be positive and the other 0-volts, one may be negative and the other 0-volts, one may be positive and the other negative, both may be positive, or both may be negative. The applications of logic to a system or a device may therefore, be further defined as follows:

1. Positive logic when the more positive potential (high) is consistently selected as the 1 state.
2. Negative logic when the less positive potential (low) is consistently selected as the 1 state.
3. Hybrid or mixed logic when both positive or both negative logic is used.

The inherent function of a logic gate is equivalent to that of a conventional switch which can be referred to as a "two-state" device and this may be illustrated by considering the theoretical circuit of a simple motor control system shown in Fig. 11.2(a). In the "off" position of the switch, the whole circuit is open, and is in an inactive or logic 0 state. In the "on" position, the switch closes the relay coil circuit causing positive d.c. to pass through the coil. Since the input voltage at A is at a high level with respect to ground then the input to the relay coil is of a logic 1 state and so the coil is energised. The input voltage at B is also high and so operates the motor from the logic 1 state existing at point C by closing the relay contacts. The circuit may therefore be considered as a positive logic function circuit.

As a further example of logic switching, let us consider the motor control circuit shown in Fig. 11.2(b). In this case, control is effected by selecting either of two parallel-connected switches located remotely from each other. If it is required to operate the motor from, say, the switch 1 location, the circuit from input A to the relay coil is closed by placing the switch at the "on" position, thereby producing an active logic 1 state in the coil circuit and at the output C. Switch 2 remains open so that the circuit from input B is in the logic 0 state. The converse would be true were the motor to be operated from the switch 2 location and with switch 1 off. The circuit is also a positive logic circuit.

GATES AND SYMBOLS
The circuits to which digital logic is applied are combinations of three basic gates performing functions referred to as "AND", "OR" and "NOT"; the last being an inverting function, and giving rise to two other gates referred to as "NAND" and "NOR".

Gate circuits are designed so that switching is carried out by either junction diodes, transistors or by a combination of both. In order to simplify diagrams as much as possible, the internal circuit arrangements are omitted, and the gates are represented by corresponding distinctively-shaped symbols which conform to accepted standards.

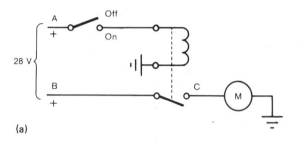

(a)

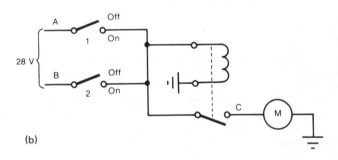

(b)

Fig 11.2
Logic switching functions

The three basic gate symbols are shown in Fig. 11.3. Variations in the symbol shapes adopted will be found in some literature, but those shown are used in the majority of manuals related to aircraft systems.

All possible combinations of input logic states and their corresponding output states, expressed in terms of the binary digits (bits) 0 and 1, can be displayed by means of truth tables. The tables appropriate to the gates referred to thus far are given in Appendix 11, and to illustrate how they are constructed, let us consider the one shown in Fig. 11.4. This corresponds to an AND gate, and as will be noted the table is a rectangular co-

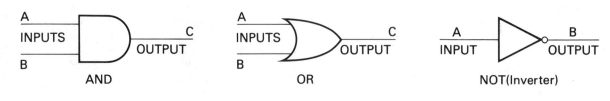

Fig 11.3
Basic gate symbols

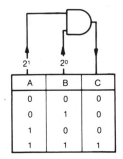

Fig 11.4
Truth table construction

ordinate presentation, the columns representing the inputs and output, and the rows representing the logic combinations.

The number of different possible combinations is expressed by 2^n, where n is the number of inputs. Thus, for a basic 2-input gate the possible combinations are $2^n = 2^2 = 4$, and so the table has two input columns and four rows. For a 3-input gate there would be eight possible combinations, and so on. The sequencing of the 0s and the 1s which make up the logic combinations, is based on identify-

ing the inputs with 2s which are raised to a power based on the input positions in a table. The table in Fig. 11.4 has two input columns, and working from right to left (this sequence always applies) column B is identified with 2^0 and column A with 2^1. Since 2 raised to zero power equals 1, one 0 and one 1 are alternately placed in each row of column B. Two raised to the power of one equals 2; therefore, two 0s and two 1s are alternately placed in column A. From column C it will be noted that an AND gate can only produce an output when the input combination is in the logic 1 state; for this reason, the gate is often referred to as an "all or nothing" gate.

The same input combinations apply to an OR gate, but as will be noted from its truth table in Appendix 11, it will produce an output when the inputs either singly, or in combination, are in the logic 1 state; the gate is therefore referred to as an "any or all" gate.

As an illustration of how gate functions may be related to theoretical circuits, let us again refer to Fig. 11.2. In order for the motor shown in diagram (a) to operate it must have a logic 1 input supplied

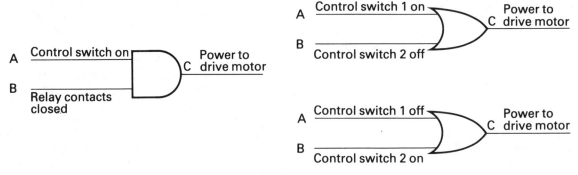

Fig 11.5
Logic gate/theoretical circuit relationship

to it, and since this can only be obtained when both the switch *and* relay contacts are closed, the circuit corresponds to an AND function and may be represented as in Fig. 11.5. For the motor in diagram (b) to operate, the logic 1 input can be supplied to it when either switch 1 *or* switch 2 is closed; thus, the circuit corresponds to an OR function and may be represented by the appropriate symbol.

The NOT logic gate is used in circuits that require the state of a signal to be changed without having a voltage at the output every time there is one at the input, or vice versa. In other words, the function of its circuit is to invert the input signal such that the output is always of the opposite state. The symbol for an inverter is the same as that adopted for an amplifier but with the addition of a small circle (called a "state indicator") drawn at either

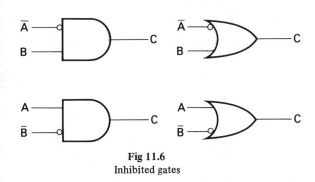

Fig 11.6
Inhibited gates

the input or output sides. When the circle is at the input side, it means the input signal must be "low" for it to be an activating signal; when at the output side, an activated output function is "low". In many cases, the NOT function is used in conjunction with the input to an AND or OR gate, as in Fig. 11.6; the gate is then said to be *inhibited* or *negated*. In order to emphasise the inversion, a line is drawn over the letter designating the inverted input. The truth tables are also given in Appendix 11, and should be compared with those of the AND and OR gates.

The addition of an inverter at the output of an AND gate and of an OR gate changes their function and they are then known respectively as NAND (a contraction of Not AND) and NOR (a contraction of Not OR) gates. They are identified by the symbols shown in Fig. 11.7.

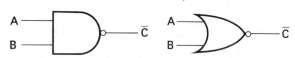

Fig 11.7
NAND and NOR gates

Figure 11.8 illustrates the symbols of two logic gates called exclusive OR and exclusive NOR, each being a combination of two inhibited AND gates and an OR gate. In some cases, the inputs to AND and OR gates may be connected together in the configuration known as "wired AND" and "wired OR". They are symbolised as shown in Appendix 11.

As an aid to the interpretation of schematic diagrams that depict the operation of systems in logic form, let us now consider some representative examples.

PRACTICAL LOGIC DIAGRAMS
Figure 11.9 relates to the operation of a twin a.c. generator system, and in particular it shows the logic states required to connect one of the generators to its respective load busbar, via the control relay (GCR) and circuit breaker (GCB). Control is effected through an OR gate, a multi-input AND gate and an inverter, and all the inputs are processed by the appropriate circuit modules of the generator control unit (GCU).

It will be noted that three of the inputs to the AND gate are negated; therefore, if the relevant circuit conditions – no internal faults within the

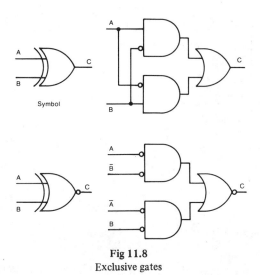

Fig 11.8
Exclusive gates

190

GCU, no faults in the bus-tie breaker (BTB) and no command signal from the external power contactor (EPC) — are satisfied, each input will be in the logic 1 state, and together with the other inputs a logic 1 output will be supplied from the AND gate to energise the "close" section of the GCB control relay. The output is also supplied to the inverter which maintains the "trip" section of the control relay de-energised under normal operating conditions.

The contacts in this section of the relay thereby complete the circuit to the GCB, which on being energised by d.c. power from the battery busbar, closes the main contacts between the generator and its load busbar. The GCB utilises magnetic latching and operates in a similar manner to that described on p. 110. With a.c. power now on the busbar, the input to the OR gate changes to logic 0, but since the output still remains logic 1 the AND gate output will be unaffected and so the GCB remains closed.

The system on which Fig. 11.9 is based is one which permits the application of power to a load

busbar from only one source at a time; in other words, sources cannot be paralleled. If, for example, external power is selected, the input F to the AND gate changes to logic 1; however, since it is inhibited the gate output will be logic 0 and so the GCB control relay and GCB itself cannot be energised. Thus, the output from an operating generator cannot be connected to the load busbar while it is being supplied from the external power source.

If any faults occur while a generator is supplying its load busbar they will be detected by the GCU, and the logic states of the appropriate inputs to the AND gate will be changed, thereby causing it to switch off outputs to the GCB control relay and to the GCB. At the same time, there will be a logic 0 input to the inverter which will energise the "trip" coil section of the relay to provide a d.c. supply to the coil of the GCB, the magnetic latch of which releases all contacts to the open position.

Figure 11.10 is an example of a logic diagram related to a system designed to give warning of low pressure in the pressurised cabin of an aircraft. In

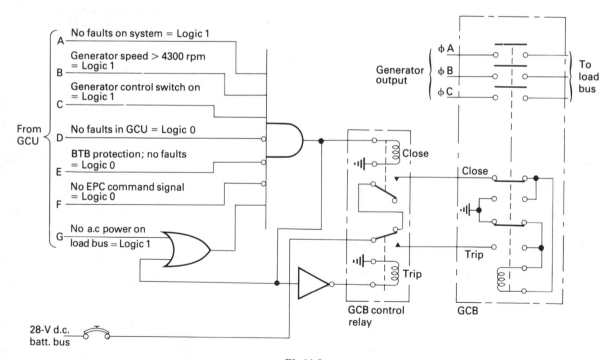

Fig 11.9
Logic diagram – generator circuit breaker control

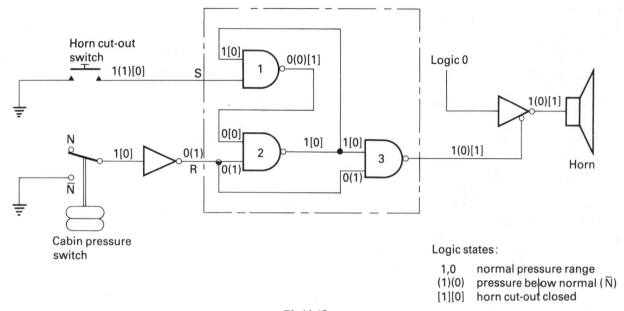

Horn cut-out switch

Logic 0

Horn

Cabin pressure switch

Logic states:

1,0	normal pressure range
(1)(0)	pressure below normal ($\overline{N}$)
[1][0]	horn cut-out closed

Fig 11.10
Logic diagram of a low-pressure warning system

this case, the logic elements used in the circuit are an inverter, a flip-flop, and a driver. A flip-flop is a bi-stable multi-vibrator device that has the basic function of storing a single bit of binary data; in this application it is comprised of three NAND gates. It is so called because the application of a suitable pulse at one input causes it to "flip" into one of its two stable states and to remain in that state, until a pulse at a second input causes it to "flop" into the other state. The driver may be considered as a form of amplifying device. Cabin pressure sensing is effected by a pressure switch which is adjusted to close under a pre-set low-pressure condition, and so provide a ground (logic 0) connection to operate the warning horn.

With the cabin pressure in its normal (N) range, the pressure switch is open, and a logic 1 is applied as input to the inverter; the inverted output is applied to NAND gate 2. The horn cut-out switch is in the open position and a logic 1 state is applied to NAND gate 1. The characteristics of the type of flip-flop used is such that with logic 0 at its reset (R) input, and logic 1 at its set (S) input, NAND gate 1 will provide a logic 0 output and apply this as the second input to gate 2. Thus, from the NAND gate truth table, gate 2 will provide a logic 1 output as a second

input to gate 1 so that it can maintain its logic 0 output. The logic 1 is also applied to gate 3, and since its second input is logic 0, it will provide a logic 1 input to the driver, the output of which is also logic 1 to maintain the warning horn in a de-activated state.

If the cabin pressure should go below normal ($\overline{N}$) it will be sensed by the pressure switch whose contacts will change over to provide a ground (logic 0) connection to the inverter. An inverted output is now applied to the R input of the flip-flop together with the logic 1 input at S; the output logic states of gates 1 and 2 therefore remain unchanged. It will, however, be noted from the diagram that the logic 1 output from the inverter is also applied as an input to NAND gate 3, so that the output to the driver now provides the logic 0 causing it to activate the horn and thereby give warning of the low cabin pressure condition.

De-activation of the horn is carried out by depressing the cut-out switch. This changes the logic input at S from 1 to 0, and since the input at R remains unchanged, the output from gate 1 is now logic 1. In tracing this through gates 2 and 3 it can be seen that a logic 1 is produced at the output of gate 3, and of the driver, so when the cabin

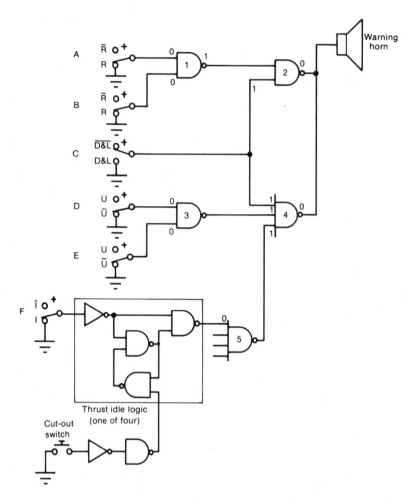

Fig 11.11
Logic diagram of a landing-gear aural warning system

pressure is below its normal range, an open-circuit condition prevails at the horn. When the cut-out switch is released a logic 1 is again applied to input S at gate 1, and because there is still a logic 1 at R, the output logic state of the flip-flop remains unchanged (or "set") until the cabin pressure switch detects that normal pressure conditions have again been attained.

Figure 11.11 illustrates the logic diagram appropriate to a system which operates a horn to warn the flight crew that the landing gear is not down and locked when the trailing edge flaps are set to the landing configuration, or when any engine thrust lever is set to the idle position. The switches A to F represent input sensors which are activated by the flaps, landing gear and engine thrust levers.

When the flaps are in the landing configuration or range (R), switches A and B are closed and so produce ground potential (logic 0) inputs to NAND gate 1; this in turn presents a logic 1 input to gate 2. If the landing gear is not down and locked ($\overline{D \& L}$) a logic 1 from switch C will also be applied as a second input to gate 2, resulting in a logic 0 output to the horn thereby causing it to sound. The logic 1 from switch C is also applied as an input to gate 4.

Since the flaps are not fully up (U) switches D and E will apply logic 0 inputs to gate 3 and this also produces a logic 1 input to gate 4. The third input to gate 4 is also logic 1 and is derived from gate 5 by inversion of the inputs produced when the corresponding thrust levers are in the idle position. The output of gate 4 remains logic 0 and the horn continues to sound.

When the landing gear is fully extended to the down and locked position, the inputs to gates 2 and 4 from the switch C will be changed to logic 0, and the horn will therefore be silenced. The horn may also be silenced by depressing the cut-out switch, thereby resetting the flip-flop in the thrust idle logic circuit.

APPENDIX ONE

Electrical and Magnetic Quantities, Definitions and Units

Quantity	Definition	Name of unit	Unit symbol	Unit definition
Electric potential	That measured by the energy of a unit positive charge at a point, expressed relative to zero potential, or earth.	Volt	V	Difference of electric potential between two points of a conductor carrying constant current of 1 ampere, when the power dissipated between these points is equal to 1 watt.
Potential difference (p.d.)	That between two points when maintained by an e.m.f., or by a current flowing through a resistance.			
Electromotive force (e.m.f.)	Difference of potential produced by sources of electrical energy which can be used to drive currents through external circuits.			
Current	The rate of flow of electric charge at a point in a circuit.	Ampere Milliampere (A x 10^{-3}) Microampere (A x 10^{-6})	A mA μA	The ampere is that constant current which, if maintained in two straight parallel conductors of infinite length, of negligible cross section, and placed 1 metre apart in vacuum, would produce between the conductors a force equal to 2 x 10^{-7} newton per metre of length.
Resistance	The tendency of a conductor to oppose the flow of current and to convert electrical energy into heat. Its magnitude depends on such factors as: nature of conductor material, its physical state, dimensions, temperature and thermal properties; frequency of current and its magnitude.	Ohm Megohm (Ω x 10^6)	Ω MΩ	The ohm is the electrical resistance between two points of a conductor when a constant p.d. of 1 volt, applied to these points, produces in the conductor a current of 1 ampere, the conductor not being the seat of any e.m.f.
Power	The rate of doing work or transforming energy.	Watt Kilowatt (W x 10^3)	W kW	Is the power which in 1 second gives rise to energy of 1 joule.
Frequency	The number of cycles in unit time.	Hertz	Hz	The definition of frequency also applies with the unit of time being taken as 1 second.

Quantity	Definition	Name of unit	Unit symbol	Unit definition
Inductance	The property of an element or circuit which, when carrying a current, is characterized by the formation of a magnetic field and the storage of magnetic energy.	Henry	H	The inductance of a closed circuit in which an e.m.f. of 1 volt is produced when the current in the circuit varies at the rate of 1 ampere per second.
Capacitance	The property of a system of conductors and insulators (a system known as a capacitor) which allows the storage of an electric charge when a p.d. exists between the conductors. In a capacitor, the conductors are known as electrodes or plates, and the insulator, which may be solid, liquid or gaseous, known as the dielectric.	Farad Microfarad ($F \times 10^{-6}$) Picofarad ($F \times 10^{-12}$)	F μF pF	The capacitance of a capacitor between the plates of which there appears a p.d. of 1 volt when it is charged by a quantity of electricity of 1 coulomb.
Electric charge	The quantity of electricity on an electrically charged body, or passing at a point in an electric circuit during a given time.	Coulomb	C	The quantity of electricity carried in 1 second by a current of 1 ampere.
Energy	The capacity for doing work.	Joule	J	The work done when the point of application of a force of 1 newton is displaced through a distance of 1 metre in the direction of the force.
Impedance	The extent to which the flow of alternating current at a given frequency is restricted, and represented by the ratio of r.m.s. values of voltage and current. Combines resistance, capacitive and inductive reactance.	Ohm	Z	
Reactance	That part of the impedance which is due to inductance or capacitance, or both, and which stores energy rather than dissipates it.	Ohm	X	
Magnetic flux	A phenomenon produced in the medium surrounding electric currents or magnets. The amount of flux through any area is measured by the quantity of electricity caused by flow in a circuit of given resistance bounding the area when this circuit is removed from the magnetic field.	Weber (Volt-second)	Wb	The magnetic flux which, linking a circuit of 1 turn, would produce in it an e.m.f. of 1 volt if it were reduced to zero at a uniform rate in 1 second.
Magnetic flux density (Magnetic induction)	The amount of magnetic flux per square centimetre, over a small area at a point in a magnetic field. The direction of the magnetic flux is at right angles to the area.	Tesla	T	Equal to 1 weber per square metre of circuit area.
Magnetic field strength (Magnetizing force)	The strength or force which produces or is associated with magnetic flux density. It is equal to the magnetomotive force per centimetre measured along the line of force.	Ampere per metre	A/m	

Quantity	Definition	Name of unit	Unit symbol	Unit definition
Magnetomotive force (m.m.f.)	The magnetic analogue of e.m.f. It represents the summated current or equivalent current, including any displacement current, which threads a closed line in a magnetic field and produces a magnetic flux along it. Can also be stated as the work done in moving a unit magnetic pole around a closed magnetic circuit.	Ampere-turns Gilbert		The product of current and the number of turns of a coil.
Reluctance	The ratio of magnetic force to magnetic flux. May be considered as the opposition to the flux established by the force. It is the reciprocal of permeance.	Ampere-turn/Weber/ Gilbert/Maxwell		
Permeability (μ)	The ratio of the magnetic flux density in a medium to the magnetizing force producing it.			
Permeance	The capability of a magnetic circuit to produce a magnetic flux under the influence of an m.m.f., and which is represented as the quotient of a given magnetic flux in the magnetic circuit and the m.m.f. required to produce it.			

Derived SI Units with Special Names

Physical quantity	Unit	Symbol	Definition of unit
Energy	joule	J	$kg\,m^2\,s^{-2}$
Force	newton	N	$kg\,m\,s^{-2} = J\,m^{-1}$
Power	watt	W	$kg\,m^2\,s^{-3} = J\,s^{-1}$
Electric charge	coulomb	C	$A\,s$
Electric p.d.	volt	V	$kg\,m^2\,s^{-3}\,A^{-1} = W\,A^{-1}$
Electric resistance	ohm	Ω	$kg\,m^2\,s^{-3}\,A^{-2} = V\,A^{-1}$
Electric capacitance	farad	F	$A^2\,s^4\,kg^{-1}\,m^{-2} = C\,V^{-1}$
Magnetic flux	weber	Wb	$kg\,m^2\,s^{-2}\,A^{-1} = V\,s$
Magnetic flux density	tesla	T	$kg\,s^{-2}\,A^{-1} = Wb\,m^{-2}$
Inductance	henry	H	$kg\,m^2\,s^{-2}\,A^{-2} = V\,s\,A^{-1}$
Luminous flux	lumen	lm	$cd\,sr$
Illumination	lux	lx	$cd\,sr\,m^{-2}$
Frequency	hertz	Hz	s^{-1}
Pressure	pascal	Pa	$kg\,m^{-1}\,s^{-2} = N\,m^{-2}$
Conductance	siemens	S	$A^2\,s^3\,kg^{-1}\,m^{-2} = \Omega^{-1}$
Viscosity, dynamic	poiseuille	Pl	$kg\,m^{-1}\,s^{-1} = N\,s\,m^{-2}$

Other Derived SI Units

Electric field strength	volt per metre	V/m	$m\,kg\,s^{-3}\,A^{-1}$
Electric charge density	coulomb per cubic metre	C/m^3	$s\,A\,m^{-3}$
Electric flux density	coulomb per square metre	C/m^2	$S\,A\,m^{-2}$
Permittivity	farad per metre	F/m	$s^4\,A^2\,m^{-3}\,kg^{-1}$
Current density	ampere per square metre	A/m^2	$A\,m^{-2}$
Magnetic field strength	ampere per metre	A/m	$A\,m^{-1}$
Permeability	henry per metre	H/m	$m\,kg\,s^{-2}\,A^{-2}$

Decimal Prefixes

Prefix	Symbol	Factor	
tera	T	10^{12}	$=1\,000\,000\,000\,000$
giga	G	10^{9}	$=1\,000\,000\,000$
mega	M	10^{6}	$=1\,000\,000$
kilo	k	10^{3}	$=1\,000$
hecto	h	10^{2}	$=100$
deca	da	10^{1}	$=10$
deci	d	10^{-1}	$=0.1$
centi	c	10^{-2}	$=0.01$
milli	m	10^{-3}	$=0.001$
micro	μ	10^{-6}	$=0.000\,001$
nano	n	10^{-9}	$=0.000\,000\,001$
pico	p	10^{-12}	$=0.000\,000\,000\,001$
femto	f	10^{-15}	$=0.000\,000\,000\,000\,001$
atto	a	10^{-18}	$=0.000\,000\,000\,000\,000\,001$

APPENDIX TWO

Ohm's Law

This law is fundamental to all direct current circuits, and can in a modified form also be applied to alternating current circuits.

The law may be stated as follows: *When current flows in a conductor, the difference in potential between the ends of the conductor, divided by the current flowing, is a constant provided there is no change in the physical condition of the conductor.*

The constant is called the resistance (R) of the conductor, and is measured in ohms (Ω). In symbols,

$$R = \frac{V}{I} \qquad (1)$$

where,

V = potential difference in volts
I = current in amperes.

Calculations involving most conductors, either singly or in a variety of combinations (see p. 199), are easily solved by this law, for if any two of the three quantities (V, I and R) are known, the third can always be found by simple transposition. Thus, from (1)

$$V = IR \text{ volts} \qquad (2)$$

$$I = \frac{V}{R} \text{ amperes} \qquad (3)$$

Power

Since some of the values used to determine the power delivered to a circuit are the same as those used in Ohm's law, it is possible to substitute Ohm's law values for equivalents in the fundamental formula for power (P) which is:

$$P = V \times I \text{ watts}$$

Thus, if $\frac{V}{R}$ is substituted for I in the power formula, it becomes

$$P = V \times \frac{V}{R} \text{ or } P = \frac{V^2}{R}$$

Similarly, if IR is substituted for V in the power formula, then

$$P = I \times I \times R \text{ or } P = I^2 R$$

By transposing the formula $P = I^2R$ to solve for the current I, we obtain

$$I^2 = \frac{P}{R}$$

from which

$$I = \sqrt{\frac{P}{R}}$$

Other transpositions of the foregoing formulae are as follows:

$$I = \frac{P}{V} \qquad V = \sqrt{PR}$$

$$R = \frac{P}{I^2}$$

$$V = \frac{P}{I}$$

APPLICATION OF OHM'S LAW TO SERIES AND PARALLEL CIRCUITS (RESISTANCES)

Circuit	Total resistance	Total voltage	Total current
 Series	$R_T = R_1 + R_2 + R_3 + \dots$ ohms or $R_T = \dfrac{V_T}{I}$ ohms If the resistances are of equal value R then: $R_T = nR$ ohms where n = number of resistors	$V_T = (I_1 R_1) + (I_2 R_2) + (I_3 R_3) +$ $\dots$ volts or $V_T = IR_T$ volts	$I_T = I_1 = I_2 = I_3 =$ $\dots$ amps or $I_T = \dfrac{V_T}{R_T}$
 Parallel	$\dfrac{1}{R_T} = \dfrac{1}{R_1} + \dfrac{1}{R_2} + \dfrac{1}{R_3} + \dots$ ohms or $R_T = \dfrac{1}{\dfrac{1}{R_1} + \dfrac{1}{R_2} + \dfrac{1}{R_3} + \dots}$ or $R_T = \dfrac{V_T}{I_T}$ If the resistances are of equal value R, then: $R_T = \dfrac{R}{n}$ When only two resistances in parallel, the total resistance is: $R_T = \dfrac{R_1 \times R_2}{R_1 + R_2}$	$V_T = V_1 = V_2 = V_3 = \dots$ volts	$I_T = I_1 + I_2 + I_3 + \dots$ amps
 Series-parallel	R_T, V_T and I_T are found by first reducing the parallel circuit to a single resistance, and then solving the whole as a simple series circuit.		

APPENDIX THREE

Power in A.C. Circuits

Real (or Average) Power

The power dissipated is $P = VI \cos \theta$ where

    V = r.m.s. voltage across circuit
    I = r.m.s. current flowing in circuit
    θ = phase angle between V and I
  $\cos \theta$ = the power factor (P.F.) of the circuit

(i) For inductors, capacitors, or circuits containing only inductors and capacitors, P.F. = 0 i.e., no power is dissipated.

(ii) For resistors and resistive circuits, P.F. = 1 i.e., power is dissipated.

(iii) For circuits containing resistance and reactance, phase angle θ varies between $0°$ and $90°$.

Real power dissipated in an a.c. circuit is also equal to I^2R and $\dfrac{V^2}{R}$ where

    I = r.m.s. current flowing in R
    V = r.m.s. voltage across R.

Reactive Power

Reactive power $P_q = VI \sin \theta$ where V, I and θ are the same as for real power.

Also $P_q = I^2X$ and $\dfrac{V^2}{X}$ where

    I = r.m.s. current in the reactance
    V = r.m.s. voltage across reactance
    X = net reactance.

P_q is measured in volt-amperes reactive (VA_r)

Apparent Power

Apparent Power $P_a = VI$ where V and I are the same as for real power.

P_a is measured in volt-amperes (VA).

APPENDIX FOUR

Connection of Capacitors and Inductors

Capacitors

IN SERIES

Capacitors in series may be considered as increasing the separation of the outer plates of the combination. Thus, the total capacitance C_T is less than the smallest capacitance of the individual capacitors, and so the relationship for C_T is similar to that for resistors in parallel, i.e.,

$$\frac{1}{C_T} = \frac{1}{C_1} + \frac{1}{C_2} + \frac{1}{C_3} + \ldots \text{ or } C_T = \frac{1}{\frac{1}{C_1} + \frac{1}{C_2} + \frac{1}{C_3} + \ldots}.$$

When only two capacitors are in series, then

$$C_T = \frac{C_1 \times C_2}{C_1 + C_2}.$$

If the capacitors are of equal value C, then $C_T = \frac{C}{n}$ where n = the number of capacitors in series.

The total working voltage rating of capacitors in series is equal to the sum of the ratings of the capacitors.

The total charge is $Q_T = Q_1 = Q_2 = Q_3 = \ldots$

IN PARALLEL

Capacitors in parallel may be considered as effectively increasing the area of the plates; therefore, since capacitance increases with plate area, the total capacitance C_T is equal to the sum of the individual capacitances. Thus, the relationship for C_T is similar to that for resistors in parallel, i.e.

$$C_T = C_1 + C_2 + C_3 + \ldots$$

The working voltage of a parallel combination is limited by the smallest working voltage of the individual capacitors.

The total charge is $Q_T = Q_1 + Q_2 + Q_3 + \ldots$

ENERGY STORED

The energy (e_C) stored in a capacitor is $e_C = \frac{CV^2}{2}$ where

    C = capacitance in farads
    V = voltage impressed across capacitor.

Inductors

Inductors in series, parallel or in combination circuits act similarly to resistors. Thus:

in series, the total inductance $L_T = L_1 + L_2 + L_3 + \ldots$

in parallel, $L_T = \dfrac{1}{\frac{1}{L_1} + \frac{1}{L_2} + \frac{1}{L_3} + \ldots}$

The energy (e_L) stored in an inductor is $e_L = \frac{LI^2}{2}$ where

    L = inductance in henrys

    I = current flow through inductor

APPENDIX FIVE

Fundamental A.C. Circuits and Formulae

Circuit	Inductive reactance (X_L) (ohms)	Impedance (Z) (ohms)	Applied voltage (V_T)	Current (I_T)
(circuit: L)	$X_L = 2\pi fL$ where f = frequency (Hz), L = inductance in henrys	As for Inductive Reactance (X_L)	$V_T = IX_L$	$I = \dfrac{V_T}{X_L}$
(circuit: L_1 L_2 series)	$X_{L\text{total}} = X_{L_1} + X_{L_2} + \ldots$		$V_T = IX_{L_1} + IX_{L_2}$	$I_T = I_{L_1} = I_{L_2}$
(circuit: L_1 L_2 parallel)	$X_{L\text{total}} = \dfrac{1}{\dfrac{1}{X_{L_1}} + \dfrac{1}{X_{L_2}} + \ldots}$		$V_T = IX_{L_1} = IX_{L_2}$	$I_T = I_{L_1} + I_{L_2}$
(circuit: C)	Capacitive reactance (X_C) Ohms $X_C = \dfrac{1}{2\pi fC}$ where f = frequency (Hz), C = capacitance (farads)		$V_T = IX_C$	$I_T = \dfrac{V_T}{X_C}$
(circuit: C_1 C_2 series)	$X_{C\text{total}} = X_{C_1} + X_{C_2} + \ldots$	As for Capacitance Reactance (X_C)	$V_T = IX_{C_1} + IX_{C_2}$	$I_T = I_{C_1} = I_{C_2}$
(circuit: C_1 C_2 parallel)	$X_{C\text{total}} = \dfrac{1}{\dfrac{1}{X_{C_1}} + \dfrac{1}{X_{C_2}} + \ldots}$		$V_T = IX_{C_1} = IX_{C_2}$	$I_T = I_{C_1} + I_{C_2}$
(circuit: R)		$Z(R) = \dfrac{V}{I}$	$V = IZ$	$I_T = \dfrac{V_T}{Z}$
(circuit: R L)		$Z = \sqrt{R^2 + X_L^2}$	$V_T = \sqrt{(IR)^2 + (IX_2)^2}$	$I = \dfrac{V_T}{Z}$
(circuit: R C)		$Z = \sqrt{R^2 + X_C^2}$	$V_T = \sqrt{(IR)^2 + (IX_C)^2}$	
(circuit: R C L)		$Z = \sqrt{R^2 + (X_L - X_C)^2}$	$V_T = \sqrt{(IR)^2 + (IX_L - IX_C)^2}$	

Resonant Circuits

From the reactance formulae given above, it will be evident that changes in frequency will change the ohmic values of reactance, e.g. a decrease in frequency decreases inductive reactance but increases capacitive reactance. At some particular frequency, known as the resonant frequency (F_n), these reactive effects will be equal, and since in a circuit containing a capacitor and inductor in series they will cancel each other, then only the ohmic value of circuit resistance would remain to oppose current flow in the circuit. Such a circuit is said to be "in resonance" and is referred to as a series resonant circuit. If the value of circuit resistance is small or consists only of the resistance in the conductors, the value of current flow can become very high. In such cases, the voltage drop across the inductor or capacitor will often be higher than the applied voltage.

Resonant frequency is determined from the formula: $F_n = \dfrac{1}{2\pi\sqrt{LC}}$. In a parallel resonant circuit, the reactances are equal and equal currents will flow through the inductor and the capacitor. Since the inductive reactance causes the current through the inductor to lag the voltage by $90°$, and the capacitive reactance causes the current through the capacitor to lead the voltage by $90°$, the two currents are $180°$ out of phase. Thus, no current would flow from the a.c. supply and the parallel combination of the inductor and the capacitor would be an infinite impedance. In practice this would not be achieved since some resistance is always present and so the parallel circuit acts as a very high impedance. The circuit is sometimes referred to as a tank circuit, or an anti-resonant circuit because of its effect being opposite to that of a series-resonant circuit.

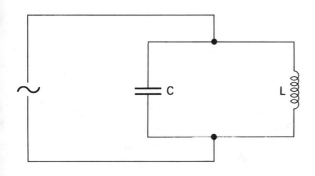

Parallel resonant circuit

APPENDIX SIX

Conversion Factors

Multiply	By	To obtain
Amperes/square metre	6.452×10^{-4}	Amperes/square inch
Amperes/square inch	1550.0	Amperes/square metre
Ampere turns/cm	2.540	Ampere turns/inch
Ampere turns/inch	0.3937	Ampere turns/cm
Ampere turns/inch	39.37	Ampere turns/metre
Btu	1054.8	Joules
Btu	2.928×10^{-4}	Kilowatt-hours
Btu/hour	0.2931	Watts
Btu/min	17.57	Watts
Circular mils	5.067×10^{-6}	Square centimetres
Circular mils	0.7854	Square mils
Circular mils	7.854×10^{-7}	Square inches
Coulombs/square metre	6.452×10^{-4}	Coulombs/square inch
Foot-pounds/min	2.260×10^{-5}	Kilowatts
Horsepower	745.7	Watts
Horsepower	0.7457	Kilowatts
Inches	1000.0	Mils
Joules	9.480×10^{-4}	Btu
Kilowatts	1.341	Horsepower
Kilowatts	56.92	Btu/min
Square centimetres	1.973×10^{5}	Circular mils
Square inch	1.273×10^{6}	Circular mils
Square mils	1.273	Circular mils
Watts	1.341×10^{-3}	Horsepower

APPENDIX SEVEN

Power Generation System Applications

Power generation systems may be classified as being
primarily either direct current or alternating current,
and from this, aircraft are generally and loosely
referred to as "d.c. aircraft" or "a.c. aircraft" depending on the power utilization requirements of components and systems. Inevitably, some systems
require power differing from that of the primary
generation system and for this purpose secondary
power conversion equipment (e.g. inverters and
TRU's) is also employed. In some isolated cases
of "d.c. aircraft", frequency-wild a.c. power is
also utilized in addition to primary d.c. generators
and power conversion equipment.

The application of these power requirements to
a representative selection of aircraft are tabulated
for reference in this Appendix on pages 206–11.

Power Generation Systems of some representative types of Aircraft

AIRCRAFT TYPE	PRIMARY POWER		D.C. GENERATOR(s)	STARTER/ GENERATOR(s)
	D.C.	A.C.		
Airbus A300	—	X	—	—
A310	—	X	—	—
A320	—	X	—	—
Aerospatiale				
ATR 42	X	—	—	X
Beech Bonanza	X	—	—	—
Baron	X	—	—	—
Duchess 76	X	—	—	—
Duke	X	—	—	—
C99	X	—	—	X
King Air	X	—	—	X
Queen Air	X	—	—	—
Starship	X	—	—	X
Boeing B707	—	X	—	—
B727	—	X	—	—
B737	—	X	—	—
B747	—	X	—	—
B757	—	X	—	—
B767	—	X	—	—
B Ae 1—11	—	X	—	—
146	—	X	—	—
Jetstream 31	X	—	X	—
125—600	X	—	X	—
700	X	—	X	—
800	X	—	—	X
748	X	—	X	—
ATP	—	X	—	—
Concorde	—	X	—	—
Viscount	X	—	X	—
Canadair				
Challenger	—	X	—	—

INVERTER(s)	ALTERNATORS			EMERG'Y CONSTANT FREQUENCY SUPPLY			REMARKS
	Self-rectifying	Variable Frequency	Constant Frequency	TRU's	APU	RAT	
X	—	—	X	X	X	—	
X	—	—	X	X	X	—	See Note 1
—	—	—	X	—	X	—	
X	—	X	—	—	—	—	See Note 2
—	X	—	—	—	—	—	
—	X	—	—	—	—	—	
—	X	—	—	—	—	—	
—	X	—	—	—	—	—	
X	—	—	—	—	—	—	
X	—	—	—	—	—	—	
—	X	—	—	—	—	—	
X	—	—	—	—	—	—	
X	—	—	X	X	—	—	
—	—	—	X	X	—	—	
X	—	—	X	X	X	—	
X	—	—	X	X	X	—	
X	—	—	X	X	X	—	
X	—	—	X	X	X	—	See Notes 1 and 3
—	—	—	X	X	—	—	
—	—	—	X	X	X	—	
X	—	—	—	—	—	—	
X	—	X	—	—	—	—	
X	—	X	—	—	—	—	See Note 4
X	—	X	—	—	—	—	
X	—	X	—	—	—	—	See Note 5
X	—	X	—	X	—	—	
—	—	—	X	—	—	—	
X	—	X	—	—	—	—	See Note 5
—	—	—	X	X	X	X	

AIRCRAFT TYPE	PRIMARY POWER		D.C. GENERATOR(s)	STARTER/ GENERATOR(s)
	D.C.	A.C.		
Cessna 150, 170, 180	X	—	—	—
210, 303, 310	X	—	—	—
340	X	—	—	—
404, 421	X	—	—	—
425, 441	X	—	—	X
500	X	—	—	X
550				
DHC -6	X	—	—	X
-7	X	—	—	X
-8	X	—	—	X
Dornier				
228	X	—	—	X
Douglas DC-9	—	X	—	—
DC-10	—	X	—	—
MD-80	—	X	—	—
Embraer Bandierante	X	—	—	X
Xingu	X	—	—	X
Brasilia	X	—	—	X
Fokker F-27	X	—	X	—
F-28	—	X	—	—
F-50	—	X	—	—
F-100				
Grumman Agcat	X	—	—	X
Gulfstream I	X	—	X	—
Gulfstream II	X	—	X	—
Gulfstream III	—	X	—	—
Lynx	X	—	—	—
Tiger	X	—	—	—
Gates Learjet 35A	X	—	X	—
Learjet 55	X	—	X	—
Lockheed				
L1011	—	X	—	—
Mooney	X	—	—	—
Maule M5	X	—	—	—

INVERTER(s)	ALTERNATORS			EMERG'Y CONSTANT FREQUENCY SUPPLY			REMARKS
	Self-rectifying	Variable Frequency	Constant Frequency	TRU's	APU	RAT	
–	X	–	–	–	–	–	
–	X	–	–	–	–	–	
–	X	–	–	–	–	–	
–	X	–	–	–	–	–	
–	–	–	–	–	–	–	
X	–	–	–	–	–	–	
					–	–	
X	–	–	–	–	–	–	
X	–	X	–	–	X	–	
X	–	X	–	X	–	–	See Note 5
X	–	–	–	–	–	–	
–	–	–	X	X	X	–	
–	–	–	X	X	X	X	
–	–	–	X	X	X	–	
X	–	–	–	–	–	–	
X	–	–	–	–	–	–	
X	–	–	–	–	–	–	
X	–	X	–	–	–	–	
–	–	–	X	X	–	–	
–	–	–	X	X	X	–	
–	–	–	–	–	–	–	
X	–	X	–	–	–	–	
X	–	X	–	X	–	–	
X	–	X	–	X	X	–	
–	X	–	–	–	–	–	
–	X	–	–	–	–	–	
X	–	–	–	–	–	–	
X	–	–	–	–	–	–	
–	–	–	X	X	X	–	
–	X	–	–	–	–	–	
–	X	–	–	–	–	–	

AIRCRAFT TYPE	PRIMARY POWER		D.C. GENERATOR(s)	STARTER/ GENERATOR(s)
	D.C.	A.C.		
Piper Apache	X	—	—	—
Aztec	X	—	—	—
Archer	X	—	—	—
Arrow	X	—	—	—
Cherokee	X	—	—	—
Cherokee Arrow	X	—	—	—
Chieftain	X	—	—	—
Mojave	X	—	—	X
Navajo	X	—	—	—
Twin Comanche	X	—	—	—
Pawnee	X	—	—	—
Pilatus B-N				
Islander	X	—	—	—
Trislander	X	—	—	—
Rockwell				
112	X	—	—	—
114 Commander	X	—	—	—
Aero Commander	X	—	—	—
Thrush	X	—	—	X
685	X	—	—	—
690B	X	—	—	X
Robin	X	—	—	—
Rallye	X	—	—	—
Saab Fairchild SF340	X	—	—	X
Shorts Skyvan	X	—	—	X
SD330	X	—	—	X
SD360	X	—	—	X
Swearingen				
Merlin IIIB	X	—	—	X

NOTE 1 Main a.c. systems non-paralleled.

NOTE 2 Alternator for heating of windshields, flight deck side windows, pitot probes.

NOTE 3 One static inverter for back-up single-phase a.c. power.

211

INVERTER(s)	ALTERNATORS			TRU's	EMERG'Y CONSTANT FREQUENCY SUPPLY		REMARKS
	Self-rectifying	Variable Frequency	Constant Frequency		APU	RAT	
	X	—	—	—	—	—	
	X	—	—	—	—	—	
	X	—	—	—	—	—	
	X	—	—	—	—	—	
	X	—	—	—	—	—	
	X	—	—	—	—	—	
	X	—	—	—	—	—	
	—	—	—	—	—	—	
	X	—	—	—	—	—	
	X	—	—	—	—	—	
	X	—	—	—	—	—	
—	X	—	—	—	—	—	
—	X	—	—	—	—	—	
—	X	—	—	—	—	—	
—	X	—	—	—	—	—	
—	X	—	—	—	—	—	
—	—	—	—	—	—	—	
—	X	—	—	—	—	—	
—	—	—	—	—	—	—	
—	X	—	—	—	—	—	
—	X	—	—	—	—	—	
X	—	—	—	—	—	—	
X	—	—	—	—	—	—	
X	—	—	—	—	—	—	
X	—	—	—	—	—	—	See Note 6
—	—	—	—	—	—	—	

NOTE 4 Alternators for windshield anti-icing.
NOTE 5 Alternator for windshield and power unit anti-icing.
NOTE 6 A.c. tapped from 3-phase windings of starter/generator for windshield anti-icing.

APPENDIX EIGHT

Electrical Diagram Symbols

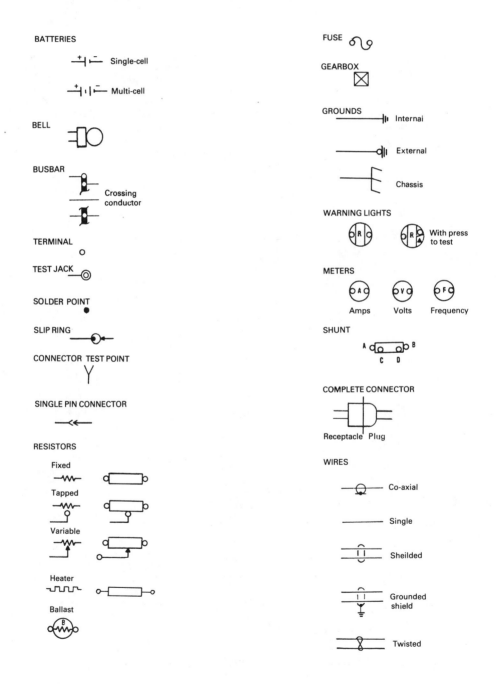

TRANSFORMERS

• No phase shift

• Phase shift 180°

Basic   Step-down   Step-up

AUTO

Fixed   Variable   Current

Wye-Wye   Wye-Delta

THERMAL DEVICES

Sensing element   Thermal resistor   Thermal relay with time delay   Continuous loop detector

N.O.
N.C.
Thermal switch

Contacts N.O.   Contacts N.C.

Thermal overload

Thermocouple

SWITCHES

S.P.S.T.   D.P.S.T.   S.P.D.T.   D.P.D.T.   Push-pull

off-Momentary ON   Push ON

Pressure

Solid-state   Reed   Push with hold-in

A(IN)   A (OUT)   A (IN)   A (OUT)

RELAYS

S.P.D.T.   S.P.S.T.   D.P.D.T. with a.c. excitation

APPENDIX NINE

Representative Aircraft Ice and Rain Protection Systems

B.A. Bleed Air; EL. Electrical; FL. Fluid; P.B. Pneumatic Boots; COMB. Combustion;

AIRCRAFT	ENGINE B.A.	ENGINE EL.	PROPELLER FL.	PROPELLER EL.	WINDSCREEN EL.	WINDSCREEN FL.	WING L.E. FL.	WING L.E. P.B.	WING L.E. B.A.	WING L.E. EL.	WING L.E. COMB.	FLIGHT CONTROLS SLATS B.A.	FLIGHT CONTROLS KR'GR FLAPS B.A.	STAB. HORIZ. P.B.	STAB. HORIZ. B.A.	STAB. HORIZ. EL.	STAB. HORIZ. FL.	STAB. HORIZ. COMB.
Airbus A300	●				●							●						
Airbus A310	●				●							●						
Boeing 707	●				●				●							●		
727	●				●							●	●					
737	●				●							●						
747	●				●													
757	●				●							●						
767	●				●							●						
BAC 1-11	●				●				●						●			
Beech 'Queen Air'			●	●	●			●						●				
Beech 99		●		●	●			●						●				
BAe 146	●				●				●						●			
Beech Baron T.C.					●			●						●				
Cessna Citation					●	●		●	●					●				
Cessna Titan					●			●						●				
Embraer Bandierante		●		●	●			●						●				
Fokker F-27		●		●	●			●						●				
F-28	●				●				●						●			
Grumman GI		●		●	●													
GII					●													
GIII					●													
BAe Jetstream 31	●			●	●			●						●				
HS 125	●				●	●	●										●	
748		●		●	●			●						●				
Falcon 50	●				●							●						
Lockheed L-1011	●				●				●									
McD DC-9	●				●							●				●		
DC-10	●				●				●									
BAC Viscount		●		●		●					●							●
Piper PA31P Navajo				●	●			●						●				
Dornier Do 228	●			●	●			●						●				

215

VERTICAL					RAM AIR SCOOPS	PITOT TUBES STATIC VENTS	WASTE WATER DRAIN	ANT'NA	AIR TEMP. PROBES	ANGLE OF ATTACK SENSORS	ICE DECT'N	RAIN	REMARKS
P.B.	B.A	EL.	FL.	COMB.	EL	EL.		EL.	EL.	EL.		EL/FL	
						●	●		●	●		●	
						●			●	●	●	●	
		●				●						●	
					●	●	●	●	●	●		●	CSD oil cooler scoops
						●	●			●		●	Outboard wing L.E. slats only heated
						●			●				
						●	●		●	●		●	
						●	●		●	●		●	
	●					●					●		Hot rod type of detector
●						●							Propellers de-iced either electrically or by fluid
●						●							
						●	●			●			Stall warning system
●						●							
●						●						●	P.B on outboard wing L.E's: El(d.c) on inboard wing L.E's. Rain repellant by B.A. FL. de-icing also used as back-up for windshield de-icing.
●						●							
	●					●							D.C. power for windshield heating
●						●				●			Stall warning system.
	●					●				●	●	●	Stall warning system. Pressure type of ice detector
						●							
						●				●			
						●				●			
						●							
●						●							
			●			●						●	FL. system as standby for windshield de-icing. Rotary 'cutter' type ice detector.
●						●							
						●							
						●		●			●	●	Vibrating type ice detector.
						●	●		●			●	
						●	●	●	●			●	Radome also de-iced.
				●		●							Exhaust gases through heat exchangers for wing de-icing
●						●							
●						●					●		Stall warning system. Propeller de-icing system utilises d.c. power.

APPENDIX TEN

Abbreviations and Acronyms associated with Electrical Systems

ACMP	AC Motor Pump	CSD	Constant Speed Drive
ACTR	Actuator	CT	Current Transformer
ACYR	Anticycling Relay	CTLR	Cross Tie Lockout Relay
AFOLT	Automatic Fire/Overheat Logic Test	CTLRS	Cross Tie Lockout Relay Slave
AMPL	Amplifier	CTR	Cross Tie Relay
ANN	Annunciator	CTTD	Cross Tie Time Delay
AP	Auxiliary Power	DAR	Drive Annunciator Relay
APB	Auxiliary Power Breaker	DBR	Dead Bus Relay
APCR	Auxiliary Power Control Relay	DBSR	Dead Bus Slave Relay
APCU	Auxiliary Power Control Unit	DDR	Drive Disconnect Relay
APGC	Auxiliary Power Generator Control	DECR	Decrease
APR	Auxiliary Power Relay	DGTL	Digital
APSR	Auxiliary Power Slave Relay	DIFF CUR	Differential Current
APU	Auxiliary Power Unit	DISC	Disconnect
AR	Annunciator Relay	DP	Differential Protection
ARR	Annunciator Reset Relay	DPCT	Differential Protection Control Transformer
ATE	Automatic Test Equipment	DPR	Differential Protection Relay
BAT	Battery	DRSR	Drive Running Slave Relay
BC	Bus Control	DSAR	Distribution System Annunciator Relay
BCHG	Battery Charger	DSBL	Disable
BCP	Bus Control Panel	DTB	Dead Tie Bus
BCTR	Battery Charger Transfer Relay	ECAM	Electronic Centralized Aircraft Monitor
BITE	Built-in Test Equipment	EDBSR	Emergency D.c. Bus Sensing Relay
BPCU	Bus Power Control Unit	E/E	Electrical/Electronics
BPP	Bus Protection Panel	ELCU	Electrical Load Control Unit
BR	Battery Relay	EMER	Emergency
BRKR	Breaker	EMI	Electromagnetic Interface
BS	Battery Switch	EP	External Power
BSR	Bus Sensing Relay	EP AVAIL	External Power Available
BTB	Bus Tie Breaker	EPC	External Power Contactor
CAC	Caution Advisory Computer	EPCR	External Power Control Relay
CB (C/B)	Circuit Breaker	EPM	External Power Monitor
CHGR	Charger	EPR	External Power Relay
CNTOR	Contactor	EPTR	Emergency Power Transfer Relay
CONT	Continuous	ERSR	Engine Running Signal Relay
CR	Control Relay	EXC	Excitation

FF	Feeder Fault		MT	Manual Trip
FFW	Reeder Fault Warning		MOT	Motor
FREQ	Frequency		MWP	Master Warning Panel
FSTD	Fault Selector Time Delay		MWS	Master Warning System
GAR	Generator Annunciator Relay		OC	Over Current
GB	Generator Breaker		OF	Over Frequency
GC	Generator Control		OFF R	Off-Reset
GCAR	Generator Control Annunciator Relay		OPP	Open Phase
GCB	Generator Control Breaker		OV	Over Voltage
GCP	Generator Control Panel		OVHT	Overheat
GCR	Generator Control Relay		OVR	Overvoltage Relay
GCR AUX	Generator Control Relay Auxiliary (contacts)		PCA	Power Control Actuator
			PCB	Printed Circuit Board
GCU	Generator Control Unit		PH	Phase
GEN	Generator		PH SEQ	Phase Sequence
GHR	Ground Handling Relay		PMG	Permanent Magnet Generator
GLR	Galley Load Relay		PMGR	Permanent Magnet Generator Relay
GND	Ground		PNL	Panel
GPCU	Ground Power Control Unit		POT	Potentiometer
GPSR	Generator Phase Sequence Relay		PRR	Power Ready Relay
GR	Generator Relay		PS	Phase Sequence
GRR	Ground Refuelling Relay		PSM	Power Supply Module
GRS	Generator Relay Slave		PSR	Phase Sequence Relay
GSAPR	Ground Service Auxiliary Power Relay		PTT	Press To Test
GSEPR	Ground Service External Power Relay		PWM	Pulse Width Modulator
GSR	Ground Service Relay		PWR	Power
GSSR	Ground Service Select Relay		RAT	Ram Air Turbine
HF	High Frequency		RCR	Reverse Current Relay
HV	High Voltage		REG (RGLTR)	Regulator
HZ	Hertz		RLY	Relay
IDG	Integrated Drive Generator		RR	Reset Relay
I/O	Input/Output		RSW	Reset Switch
INOP	Inoperative		RVDT	Rotary Variable Displacement Transformer
INTL	Interlock		SCR	Silicon Control Rectifier
INV	Inverter		SEL	Selector
ISB	Inter System Bus		SEQ	Sequence
LCD	Liquid Crystal Display		SGU	Symbol Generator Unit
LED	Light Emitting Diode		SHLD	Shield
LRU	Line Replaceable Unit		SNSR	Sensor
LT	Light		SOL	Solenoid
LV	Low Voltage		SOLV	Solenoid Valve
LVDT	Linear Variable Displacement Transformer		SPLY	Supply
MCDF	Maintenance Control & Display Panel		SR	Starter Relay
MC & W	Master Caution & Warning		SSR	Solid State Relay

218

STBY	Standby	UF	Under Frequency
SW	Switch	UFR	Under Frequency Relay
SYNC	Synchronise	UFTD	Under Frequency Time Delay
TB	Tie Bus	UNBAL	Unbalance
TBDP	Tie Bus Differential Protection	UV	Under Voltage
TBF	Tie Bus Fault	UVR	Under Voltage Relay
TBSR	Transfer Bus Sensing Relay	UVTD	Under Voltage Time Delay
TD	Time Delay	VR	Voltage Regulator
TMR	Timer	VRAR	Voltage Regulator Annunciator Relay
TP	Test Point	WLDP	Warning Light Display Panel
TR (T-R)	Transformer Rectifier	XDCR	Transducer
TRCR	Transfer (Bus) Control Relay	XFMR	Transformer
TRU	Transformer Rectifier Unit	XFR	Transfer
UBR	Utility Bus Relay	XMTR	Transmitter

APPENDIX ELEVEN

Logic Gates and Truth Tables

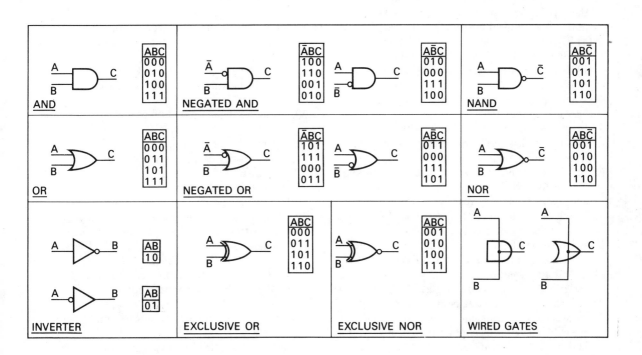

Exercises

Chapter 1

1. Describe how direct current is produced by a generator.
2. Describe how generators are classified, naming the three classes recognized and the class normally employed in aircraft systems.
3. (a) Briefly describe armature reaction and the effects it has on generator operation.
   (b) How is armature reaction corrected in aircraft generators?
4. What is meant by reactance sparking? Explain how it is counteracted.
5. In connection with generator brushes, state:
   (a) the materials from which they are made;
   (b) why several pairs of brushes are used.
6. Briefly describe the causes of brush wear under high altitude flight conditions and the methods adopted for reducing wear.
7. Which of the factors affecting the output voltage of a generator is normally controlled?
8. With the aid of a circuit diagram, describe the fundamental principle of the carbon pile method of voltage regulation.
9. Describe how the voltage output of a generator is controlled by a vibrating contact type of regulator.
10. How is the generator shunt-field resistance controlled by a vibrating contact type of regulator under heavy external load conditions?
11. What additions must be made to voltage regulation circuits of a multi-generator system?
12. With the aid of a circuit diagram describe how parallel operation of generators can be obtained.
13. Describe a means for cooling aircraft generators.
14. Briefly describe how the d.c. power is derived in aircraft utilizing a frequency-wild alternator system.
15. What principal methods are used for the driving of generators?
16. What are the typical contact arrangements of transistors? Describe how current is made to flow in one of these arrangements.
17. State the functions of a Zener diode in the circuit of a solid-state type of voltage regulator.
18. What are the principal functions of batteries in aircraft?
19. Describe the construction of a lead-acid battery and the chemical changes which occur during charging.
20. Describe the construction of a nickel-cadmium battery and the chemical changes which occur during charging.
21. The capacity of a battery is measured in:
    (a) volts.
    (b) cubic centimetres.
    (c) ampere-hours.
22. What indications would be displayed by a lead-acid battery of the free electrolyte type, and a nickel-cadmium battery, which would serve as a guide to their state of charge?
23. Describe a typical method of extracting fumes and gases from the battery compartment of an aircraft.
24. What do you understand by the term "thermal runaway"?
25. What is the purpose of using a parallel-to-series configuration of batteries in some types of aircraft?
26. With the aid of a circuit diagram, describe how in some types of aircraft the battery may be charged from an external power unit.
27. How are batteries which are installed in most types of large transport aircraft maintained in a state of charge?

Chapter 2

1. The frequency of an alternator may be determined by:
   (a) dividing the number of phases by the voltage.
   (b) multiplying the number of poles by 60 and dividing by the rev/min.
   (c) multiplying the rev/min by the number of pairs of poles and dividing by 60.
2. Explain the term r.m.s. value.
3. The current in a purely capacitive circuit will:
   (a) lead the applied voltage.
   (b) lag the applied voltage.
   (c) be in phase with the applied voltage.
4. (a) With the aid of circuit diagrams briefly describe the two methods of interconnecting phases.
   (b) State the mathematical expressions for calculating line voltage and line current in each case.
5. The phase voltage of a three-phase star-connected a.c. generator is:
   (a) equal to line voltage.
   (b) greater than line voltage.
   (c) less than line voltage.
6. The ratio of true power to apparent power of an a.c. circuit is known as:
   (a) reactive power.
   (b) power factor.
   (c) real power.
7. The impedance of an a.c. circuit is measured in:
   (a) kilovolt-amperes.
   (b) amperes.
   (c) ohms.
8. How is Power Factor affected by a circuit containing inductance and capacitance?
9. What do you understand by the term "frequency-wild system"?
10. State the factors upon which the frequency output of an a.c. generator depend.
11. For what type of load is a frequency-wild supply most suitable?
12. With the aid of a schematic diagram, describe how a frequency-wild generator can be excited and how its output voltage can be controlled.
13. In a CSD unit, the control cylinder is mechanically coupled to the:
    (a) variable displacement unit.
    (b) fixed displacement unit.
    (c) governor.

14. The governor of a CSD unit is driven by the:
    (a) input gear.
    (b) input ring gear.
    (c) output gear shaft.
15. In which phase is a CSD unit said to be operating when the governor causes charge oil to flow into the control valve?
16. Explain the purpose of the "fine" adjustment of the governor, and how it is accomplished.
17. A CSD unit which has been disconnected during flight:
    (a) may only be reconnected on the ground.
    (b) may be reconnected in flight by re-set mechanisms.
    (c) automatically resets at engine shutdown.
18. What is the purpose of the a.c. exciter and rotating rectifier assemblies of a constant-frequency generator?
19. Explain how temperature effects on an a.c. exciter are compensated.
20. What factors must be controlled when constant-frequency a.c. generators are operated in parallel?
21. What is the meaning of kVAR and to which of the factors does it refer?
22. When constant-frequency generators are operating in parallel, the sharing of real load is controlled by:
    (a) varying the excitation current in each generator.
    (b) varying the output speed of the CSD units.
    (c) shedding certain loads.
23. State the functions which current transformers can perform in controlling load sharing between constant-frequency generators.
24. What is a mutual reactor and in which section of a load-sharing circuit is it used?

Chapter 3

1. Rectification is the process of converting:
   (a) a high value of a.c. into a lower value.
   (b) d.c. into a.c.
   (c) a.c. into d.c.
2. Describe the fundamental principle on which rectification is based.
3. An "n-type" semiconductor element is one having:

(a) an excess of "holes".

(b) a deficiency of "holes".

(c) an excess of electrons.

4. What semi-conductor elements are usually employed in rectifiers used in aircraft? Describe the construction of one of these rectifiers.

5. What is meant by the term Zener voltage?

6. Is the Zener voltage of any practical value in rectification equipment?

7. Explain the operating principle of a silicon-controlled rectifier (S.C.R.).

8. With the aid of a circuit diagram explain how full-wave rectification of a three-phase input takes place.

9. Describe the basic construction and principle of the device used for converting alternating current from one value to another.

10. What is meant by transformation ratio and how is it applied to "step-up" and "step-down" transformers?

11. Draw a circuit diagram to illustrate a star-connected three-phase transformer.

12. Describe the operation of a current transformer. For what purpose is such a device used?

13. What effects do changes in frequency have on the operation of a transformer?

14. With the aid of a circuit diagram, describe the operating principle of a typical transformer-rectifier unit.

15. For what purpose are the power converting machines of the rotary type utilized in aircraft?

16. Describe a method of regulating the voltage and frequency of a rotary inverter.

17. Describe how transistors are utilized for the conversion of electrical power supplies and regulation of voltage and frequency levels.

Chapter 4

1. Explain why it is necessary for an external power supply circuit to form part of an aircraft's electrical system.

2. Draw a diagram of a basic external d.c. power supply circuit and explain its operation.

3. In a multi-pin plug how is it ensured that the breaking of the ground power supply circuit takes place without arcing?

4. Draw a diagram of an external power supply circuit of a typical "all-a.c." aircraft and explain its operation.

5. What principal items are located on a typical control panel as provided on some types of aircraft?

6. State the purpose of an APU and the services usually provided by it.

Chapter 5

1. What is the function of busbars and what form do they normally take?

2. What is meant by a split busbar system?

3. Define the three groups which usually categorize the importance of consumer services.

4. State the function of a bus-tie breaker and the type of busbar arrangement to which it would be applied.

5. Describe three different types of electrical cable commonly used in aircraft, stating their properties, limitations and identifications. State a typical application of each type.

6. What principal methods are adopted for routing cables through an aircraft?

7. Describe a method of routing wires and cables from a pressurized to a non-pressurized area of an aircraft.

8. Name some of the materials used for thermo-couple cables and state their applications.

9. What is meant by earthing or grounding?

10. How is a ground system formed in an aircraft the primary structure of which is non-metallic?

11. What is a crimped terminal?

12. What is the function of an in-line connector?

13. What precautions must be taken when making aluminium cable connections?

14. How is it ensured that a plug mates correctly with its socket?

15. State how plug pins and socket cavities are identified and also how their sequencing is signified.

16. Discuss briefly the process of "potting" a cable to a plug or socket.

17. What are the principal functions of a bonding system?

18. State some of the applications of primary and secondary bonding.

19. Briefly describe the methods generally adopted for the discharging of static.

20. What is the purpose of screening?

Chapter 6

1. The number of circuits which can be completed through the poles of a switch is indicated by the term:
   (a) pole.
   (b) position.
   (c) throw.
2. What do you understand by the term "position" in relation to toggle switches?
3. To which circuits are (a) push-switches and (b) rotary switches normally applied?
4. Describe the construction and operation of a micro-switch.
5. What are the three main stages of movement of a micro-switch operating plunger?
6. Describe the construction and operation of a mercury switch arranged to "break" a circuit.
7. In a thermal switch employing steel and invar elements, actuation of the contacts under increasing temperature conditions is caused by:
   (a) expansion of the steel element only.
   (b) contraction of the invar element only.
   (c) expansion of the steel element causing displacement of the invar element.
8. What are the principal ways in which relays may be classified?
9. What do you understand by the terms "pull-in" voltage and "drop-out" voltage?
10. Sketch a cross-section of a typical pressure switch; explain its operation.
11. What type of relay is required for a circuit in which control circuit current is of a very low value? Briefly describe the relay and its operation.
12. (a) For what purpose are "slugged" relays used?
    (b) Describe the methods usually adopted for obtaining the slugging effects.
13. Explain how the contacts of a typical bus-tie breaker remain in the latched position.

Chapter 7

1. What are the principal differences between a fuse and a current limiter as far as functions and applications are concerned?
2. State the function of a limiting resistor, and with the aid of a circuit diagram describe a typical application.

3. A circuit breaker is a device for:
   (a) protecting an electrical circuit from current overload;
   (b) collapsing the primary circuit of a magneto;
   (c) completing a circuit without being affected by current flow.
4. With the aid of a sketch, describe the construction and explain the principle of operation, and characteristics, of a thermal circuit breaker.
5. What is meant by the term "trip free" when applied to a thermal circuit breaker?
6. Under what conditions would you say that it is permissible for a circuit breaker to be used as a switch?
7. What do you understand by the term "reverse current"?
8. Describe the operation of a reverse current cut-out relay.
9. What is the function of a reverse current circuit breaker?
10. Briefly describe the operating principle of a reverse current circuit breaker.
11. Describe a typical method of protecting a d.c. generating system against overvoltage.
12. What is the purpose of incorporating time delays in the undervoltage and overvoltage protection circuits of constant frequency generating systems?
13. What are the overall effects of over-excitation and under-excitation on a.c. busbar voltage, and how is protection provided?
14. What is meant by a differential or feeder fault and how is it caused?
15. Briefly describe the operation of a differential current protection system.

Chapter 8

1. Describe the operating principle of a moving coil instrument.
2. Can moving coil instruments be directly connected in the circuits of a.c. systems for measurement of voltage and current, or is it necessary for them to be used with certain other components?
3. A soft-iron core is placed within the coil of a moving coil instrument because:
   (a) it provides a solid spindle about which the coil can rotate.
   (b) this ensures an even, radial and intensified magnetic field for the coil to move in.

(c) the inertia of the core will damp out oscillations of the coil and pointer.

4. Describe how ammeters can measure very high current values without actually carrying full load current.

5. How are moving coil instruments protected against the effects of external magnetic fields?

6. What is the purpose of central warning systems? Briefly describe a typical system.

7. Define the acronyms ECAM and EICAS.

8. What is the format of the displays presented by the display units of the ECAM system?

Chapter 9

1. Define the characteristics which govern the application of a d.c. motor to a particular function.

2. What are the principal characteristics of a shunt-wound and series-wound motor?

3. When the r.p.m. of a shunt-wound motor increases the current drawn by it:
   (a) decreases.
   (b) remains the same.
   (c) increases.

4. Draw a circuit diagram of the motor to be applied to a system where high starting torque and steady "off-load" running is required.

5. What is meant by the term "shunt limiting"?

6. With the aid of a circuit diagram explain the operation of a motor required for simple reversing functions.

7. Actuator motors are prevented from over-running their limits of travel by means of:
   (a) manually controlled switches.
   (b) electromagnetic brakes.
   (c) cam-operated limit switches.

8. (a) Explain how a three-phase rotating magnetic field is produced in an induction motor.
   (b) Why does the rotor run at a speed slightly less than that of the rotating field?

9. In an a.c. motor, the difference between synchronous speed and the speed of the rotor is termed:
   (a) the motor loss speed.
   (b) the brake speed.
   (c) the slip speed.

10. What is the formula for determining the synchronous speed of an induction motor?

11. In terms of the amount of field rotation relative to one cycle of the power supply, what are the differences between 2-pole, 4-pole and 6-pole motors?

12. Describe how a rotating magnetic field is produced in a single-phase induction motor.

13. Describe the operation of a hysteresis motor and state one of its applications.

Chapter 10

1. A typical frequency of anti-collision light beam rotation is:
   (a) 40–45 cycles per minute.
   (b) 80–90 cycles per second.
   (c) 80–90 cycles per minute.

2. Why are the two surfaces of a V-shaped reflector arranged differently from each other?

3. What are the principal functions of a strobe lighting system?

4. Describe the operating principle of a strobe lighting system.

5. Give a brief description of the construction of a typical landing light unit.

6. At which stage of landing light operation is the light normally illuminated?

7. How is it ensured that the beam of an extending/retracting type of landing light located in a flap track, remains parallel to a fore and aft datum during flap extension?

8. By means of a diagram show the interconnection of the components of a simple engine starting system.

9. In terms of cranking speeds what are the differences between starter motor requirements for reciprocating and turbine engines?

10. The self-sustaining speed is the:
    (a) maximum speed at which the starter motor runs to maintain rotation of an engine.
    (b) speed at which the engine is capable of maintaining rotation.
    (c) speed at which current to the motor is interrupted.

11. What type of motor is used for engine starting purposes?

12. What is the function of an overspeed relay fitted in some turbine engine starting systems? Describe how it fulfils this function.

13. The purpose of a "blow-out" cycle is to:
    (a) remove excess air from an engine during starting.
    (b) blow cooling air through the starter motor after starting.
    (c) remove the unburnt fuel from an engine in the event of an unsuccessful start.
14. In systems incorporating a "blow-out" facility, why is it necessary for the motor running time to be limited?
15. Describe the operation of a typical starter-generator system.
16. The contact breaker of a magneto is connected in the:
    (a) primary winding circuit.
    (b) secondary winding circuit.
    (c) circuit between distributor and spark plugs.
17. Explain how the rate of collapsing of the primary winding flux is increased.
18. What is the formula for calculating the speed of a magneto?
19. A rotating armature magneto to be fitted to a 6-cylinder engine must be driven at:
    (a) the same speed as the engine.
    (b) half the speed of the engine.
    (c) one and a half times the engine speed.
20. Why is it necessary for the output of a magneto to be boosted during starting? Describe a method of achieving this.
21. What are the essential differences between low and high tension magneto systems?
22. What are the materials generally used for the insulators and electrodes of spark plugs?
23. What are the essential differences between a turbine engine ignition system and the system used for a reciprocating engine?
24. With the aid of a circuit diagram, explain the operation of a high energy ignition unit.
25. What is the purpose of a "relight" circuit and what methods are adopted?
26. Describe the construction of a "firewire" type of detecting element, and state the effects that temperature changes have on it.
27. In what way does a detecting element of the Systron Donner System differ from a "firewire" type?
28. Describe the operation of a typical smoke detector.
29. Describe the operation of an electrically-operated fire extinguisher.

30. Describe the two techniques "de-icing" and "anti-icing".
31. What types of electrical heating elements are used for the de-icing of propellers and engine air intakes?
32. How is electrical power transmitted to propeller blade heating elements?
33. What types of heating elements and power supplies are used for the anti-icing of windshields?
34. In a propeller and air intake system using power cycling, to what air temperatures do "fast" and "slow" selections correspond?
35. In what sequence is current supplied to the heater elements of a propeller de-icing system which is d.c. operated?
36. Describe the operation of a control method adopted in a typical windshield anti-icing system.
37. For what purpose is a.c. and d.c. power utilized in hot-air bleed anti-icing systems?
38. State the purpose of the anticipator and limit sensors in the ducting of an airconditioning system.
39. How is a desired cabin temperature signal established when an airconditioning system is operating in "auto"?
40. How is a mix valve prevented from staying at a "too hot" position?
41. In the event of power control being lost, how is the airconditioning ducting protected against overheat?
42. Describe the operation of a propeller synchronizer system.

Chapter 11

1. Name the functions performed by the three basic logic gates.
2. The logic symbol shown in Fig. E11.1 represents:

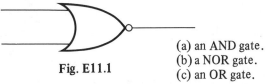

Fig. E11.1

(a) an AND gate.
(b) a NOR gate.
(c) an OR gate.

3. In order to energise the relay in the circuit shown in Fig. E11.2, the logic state at the inputs must be:

226

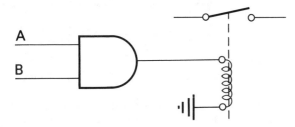

Fig. E11.2

(a) logic 0 at points A and B.
(b) 0 at point A and 1 at point B.
(c) 1 at both points.
4. What is the significance of the small circle drawn at the inputs or output of some types of logic gates?
5. What is the purpose of logic gate truth tables?
6. The truth table shown in Fig. E11.3 corresponds to:

A	B	C
1	1	1
1	0	1
0	1	1
0	0	0

(a) an OR gate.
(b) an AND gate.
(c) a NOR gate.

Fig. E11.3

7. The circuit shown in Fig. E11.4 performs a logic:

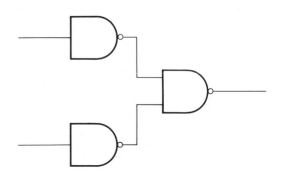

Fig. E11.4

(a) AND function.
(b) OR function.
(c) NAND function.
8. What is the significance of the line drawn over a letter or signal function when related to the input or output of a logic gate?

Solutions to Exercises

Chapter 1	21	(c)		Chapter 8	3	(b)
Chapter 2	1	(c)		Chapter 9	3	(b)
	3	(a)			7	(c)
	5	(c)			9	(c)
	6	(b)				
	7	(c)		Chapter 10	1	(b)
	13	(a)			10	(b)
	14	(c)			13	(c)
	17	(a)			16	(a)
	22	(b)			19	(b)
Chapter 3	1	(c)		Chapter 11	2	(b)
	3	(c)			3	(c)
					6	(a)
Chapter 6	1	(c)			7	(b)
	7	(a)				
Chapter 7	3	(a)				

228

Index